JOURNAL OF SCIENTIFIC EXPLORATION
A Publication of the Society for Scientific Exploration

(ISSN 0892-3310)

Editorial Office: *Journal of Scientific Exploration*, Society for Scientific Exploration, Kathleen E. Erickson, *JSE* Managing Editor, 151 Petaluma Blvd. So., #227, Petaluma, CA 94952 USA EricksonEditorial@gmail.com, 1-415-435-1604, (*fax* 1-707-559-5030)

Manuscript Submission: Submit manuscripts online at http://journalofscientificexploration.org/index.php/jse/login

SUBSCRIPTIONS & PREVIOUS JOURNAL ISSUES: Order forms on back pages or at scientificexploration.org.

Journal of Scientific Exploration (ISSN 0892-3310) is published quarterly in March, June, September, and December by the Society for Scientific Exploration, 151 Petaluma Blvd. So., #227, Petaluma, CA 94952 USA. Society Members receive online *Journal* subscriptions with their membership. Online Library subscriptions are $165.

JOURNAL OF SCIENTIFIC EXPLORATION
A Publication of the Society for Scientific Exploration

AIMS AND SCOPE: The *Journal of Scientific Exploration* publishes material consistent with the Society's mission: to provide a professional forum for critical discussion of topics that are for various reasons ignored or studied inadequately within mainstream science, and to promote improved understanding of social and intellectual factors that limit the scope of scientific inquiry. Topics of interest cover a wide spectrum, ranging from apparent anomalies in well-established disciplines to paradoxical phenomena that seem to belong to no established discipline, as well as philosophical issues about the connections among disciplines. The *Journal* publishes research articles, review articles, essays, commentaries, guest editorials, historical perspectives, obituaries, book reviews, and letters or commentaries pertaining to previously published material.

JOURNAL OF SCIENTIFIC EXPLORATION
A Publication of the Society for Scientific Exploration

Volume 26, Number 4   2012

Articles of Interest

SSE News

Journal of Scientific Exploration, Vol. 26, No. 4, pp. 763–766, 2012   0892-3310/12

Recently, I was chosen to be one of forty individuals invited to submit a short essay for a special issue of the *Journal of Parapsychology*. The topic we were asked to address was "Where will parapsychology be in the next 25 years?" This challenge encouraged me to reflect on the history of psi research and the continuing debate over its merits and results, and I found it to be a useful exercise. I also found it to be somewhat depressing, and I don't believe that was due merely to an uncharacteristic spasm of negativity. In any case, it led me to wonder what my SSE colleagues working in other areas of anomalies research would say about the future of their own respective disciplines. For example, can we speculate competently, based on the history of UFO research, where that research is likely to be in a quarter century? Or the quest for new and reliable sources of energy, including "cold fusion" or LENR? Hypnosis or altered states research? Are we seeing any notable and sweeping advances in any of these areas—say, increases in understanding among the few who do the research, or in the impact that research is having in the public at large? Can we predict competently which of the latest cutting-edge or trendy theoretical proposals (e.g., in earth science, astrophysics, or survival research) are likely to be genuinely fruitful?

So, in the hope that my reflections on the future of parapsychology will encourage similar exercises among *JSE* readers (in addition to eliciting predictable cries of protest), what follows is an expanded and (freed from severe space limitations) rather less curmudgeonly version of my little essay for the *Journal of Parapsychology*, reproduced in part with the kind permission of editor John Palmer. Perhaps it would be interesting to have a roundtable discussion along these lines at some future SSE conference.

■ ■ ■

I don't believe I'm a pessimistic person, but I find it difficult to be optimistic about the next 25 years of parapsychological research. That's because when I consider the field's successes and failures since the late nineteenth century, certain patterns stand out starkly for me.

First, skepticism about the reality of ESP, psychokinesis, or the evidence for postmortem survival, has always been intense, especially in scholarly circles, and it has quite often been vicious, recalcitrant, and dishonest. Granted, over the years, some open-minded scientists and others have dispassionately (or otherwise) reviewed the evidence and found themselves persuaded either about the reality of the phenomena or at least the value of doing additional research. But these people clearly comprise a very small minority, and parapsychologists have clung anxiously to their most prominent members in order to tout their endorsements or support. To take just one example: we're too often reminded that Brian Josephson—who does no psi research but who actively and effectively defends it—is a Nobel-winning physicist who takes parapsychological research seriously. Don't misunderstand me; I also welcome Brian's vigorous support, his willingness to contribute to the theoretical dialogue in parapsychology, and his many—time consuming—efforts to combat shoddy skepticism. But personally, I'm embarrassed by parapsychologists' frequent and dialectically shabby appeals to his authority and prestige. Partisans from all sides of the psi debate can point to prominent scientists who share their points of view, and respected, intelligent people can say foolish things. It's both pathetic and irrelevant—and it does appear desperate—to cite Brian's (or any supporter's) credentials. What matters is what they say.

The fact is, the resistance to the entire field of psi research hasn't diminished significantly in more than a century, and the tactics employed to discredit the field or its major figures have remained the same as well. Critics have all along feigned certitude about the worthlessness of the data while betraying their ignorance of what the data actually are. Detractors (or deniers) still employ fallacious argumentative strategies (e.g., ad hominem or straw-man arguments) they would be quick to detect and denounce if they had been the targets of those arguments instead. And not surprisingly, the tone of these criticisms often reveals an intensity of emotion inappropriate to what should be an open-minded empirical inquiry. Indeed, it looks conspicuously like a fear response. A contributing factor, of course, is that somewhere along the line (but undoubtedly beginning many centuries ago), scientists and scholars allowed ego, pride, or self-interest to interfere with the thrill of discovery. Why aren't more of us excited to learn that we might have been wrong, so long as the discovery brings us closer to getting things right?

Second, it's clear that parapsychology's gradual adoption of more relentlessly and sophisticated quantitative methods of investigation has made almost no difference to the course of skeptical opposition. On the contrary, it's simply opened a new and fruitful—and largely technical and specialized—playing field for glib or dishonest criticism. So instead of concentrating on allegations of mediumistic fraud, biased observation, sloppy reporting, or faulty memory, critics now focus (for example) on allegedly questionable statistics, the proper criteria for conducting meta-analyses, or other methodological flaws (real or imagined). In that respect, J. B. Rhine's so-called "revolution" of moving from mediumistic case studies to quantitative lab experiments has been a complete failure. Overall, neither the public at large nor the subset of academic detractors has been more convinced by quantitative research than they were beforehand by anecdotal reports and mediumistic case studies.

Of course, all sides in the psi debate (believers, doubters, and deniers) are guided by some combination of intuition (or "passion") and reason. Nevertheless, spontaneous case studies have always been, and continue to be, more impactful—and in important ways more clear-cut—than a study whose conclusions rely on controversial and very abstract reasoning, either about statistical presuppositions, quantum weirdness, or the nature of causality. Significantly, not even all psi researchers consider themselves convinced about the reality of the phenomena they study, and I believe it's true that most of the doubters (or fence-sitters) assume that conviction can only come from applying some version of probability theory to lab experiments and from determining ostensible paranormality solely by the numbers.

Now if only there were a growing or robust trend in current parapsychology to focus more on field work or exceptional subjects, and to try to get a handle on the phenomena's role in life, there might be reason to think we're finally starting to get somewhere. We might then have a better idea of what it is, exactly, we're trying to study experimentally—not to mention whether (or to what extent) experimental methods, confidence intervals, and p-values are even appropriate to the phenomena. And then maybe we'll have a better grip on how to elicit the phenomena with greater reliability (or at least a better understanding of why the phenomena are doomed to remain somewhat elusive). But that's not happening, and overall the dialogue between critics and defenders of psi research (and to some extent the conversation among parapsychologists themselves) hasn't budged significantly in many decades. It continues to center too often on difficult-to-resolve alleged methodological shortcomings or statistical errors—at least when critics aren't merely flaunting their ignorance of the data or else echoing the old skeptical mantra about the supposedly intuitively obvious impossibility of the phenomena.

Actually, some critics (and even a few seasoned parapsychologists) continue to make the more egregious error of thinking that we first need a well worked-out *theory* before we can admit that the phenomena are real. But of course it's completely obvious that we can know *that* something is the case without knowing *why* it's the case. Yet this flawed objection, like the other lame critical strategies mentioned earlier, shows no signs of disappearing from the field of debate.

Another discouraging trend (revealed even more clearly from my privileged perspective as a journal editor) is that too many people publish (or try to publish) books and articles about parapsychology (pro and con), or conduct their own experiments, with little or no grounding in the field's extensive literature, both empirical and theoretical. In my editorial for *JSE* 23(3), I lamented, for example, how presenters at SSE conferences, with a solid background in conventional scientific research, try to conduct parapsychology experiments (say in healing) in apparent ignorance of two related and well-known methodological problems: (a) the "source of psi" problem about the extent to which a controlled experiment is even possible in parapsychology, and (b) well-documented experimenter expectancy effects in behavioral research. It should be obvious that a background in some mainstream scientific discipline is not, *by itself*, qualification for publishing opinions about parapsychology or conducting one's own experiments. But this form of naïveté—if not outright hubris—is regrettably quite common, and I encounter it repeatedly in manuscripts submitted to the *JSE*. Of course, as a result, simplistic and ignorant opinions (pro and con) about psi research spread and perpetuate. This can only impede the dissemination and recognition of sensible and informed work in the area.

I know some will disagree with my bleak assessment and point to apparent inroads here and there within the scientific community. But of course there have been scattered successes. Some formerly intransigent skeptics have adopted more moderate positions; some who had previously opposed all things paranormal now display degrees of sympathy or support; and occasionally a paper on psi research appears in a respectable mainstream journal (usually accompanied and followed by a chorus of outrage). But that's always been the case, and I'm still awaiting evidence suggesting that the optimists have identified a lasting trend and aren't simply ignorant of the field's history or otherwise empirically myopic, or (equally likely) inductively challenged. In the meantime, funding remains scarce and modest, educational opportunities and stable research positions are few and far between, and the academy remains a generally hostile environment.

I'm not saying this will *never* change. After all, I do believe in the inexorable (though not smooth or steady) advance of human knowledge, and I'm actually confident that humankind (if it persists long enough) will eventually progress to points at which psi phenomena are generally accepted as real and in which the phenomena get incorporated into one or more widely established conceptual frameworks. But this will be a huge and deep change. After all, people don't relinquish old habits and entrenched beliefs without a real struggle, and in fact it's very difficult to reason people out of positions they haven't been reasoned into. So I wouldn't bet on major progress or success in parapsychology happening any time soon.

Now as far as I'm concerned, that doesn't mean that researchers should throw in the towel, or that the *JSE* should stop publishing good quantitative (or qualitative) papers in parapsychology. Although I don't think we can look forward to rapid progress, there's always room for good research (both in the lab and in the field), and in fact there's a continuing *need* to accumulate such research. That's the only way the truth will out. So I'm pleased that this journal and the SSE can play a vital role in the process, and I'm proud that the *JSE* so consistently delivers research and essays of the highest quality in many areas of frontier science.   **Stephen E. Braude**

Journal of Scientific Exploration, Vol. 26, No. 4, pp. 767–779, 2012   0892-3310/12

RESEARCH ARTICLE

The Bell Inequality and Nonlocal Causality

CHARLES W. LEAR

Physics Department, Weber State University, Ogden, UT 84403
charleslear@weber.edu

Submitted 8/2/2011, Accepted 6/20/2012

Abstract—The Freedman–Clauser experiment in Berkeley and the Aspect experiment in Orsay were the defining physical experiments demonstrating nonlocal causality in quantum mechanics. They each counted coincidence measurements on entangled polarized photons from a common source. This article begins with a brief discussion of the quantum mechanics of polarized photons. We show an example of the changes in the count rates when the polarizers are changed under assumptions of local causality. This causes a contradiction with quantum mechanical predictions. The example uses a logical flow and the algebra of inequalities. It constitutes a conditional proof of the Bell inequality. Next we discuss the experimental background and the events leading up to it. We discuss several hypotheses in explanation, of which our favored is the time reversal of cause and effect.

Keywords: Bell's theorem—causality—entanglement—nonlocality—time reversal

Introduction

The goal of this article is to provide an accessible description of Bell's inequality and the physical basis of nonlocality. Bell's theorem (Bell 1964) is a mathematical inequality that proves that under some conditions quantum mechanics is inconsistent with local causality. What is local causality? Figure 1 is a spacetime diagram with time increasing horizontally to the right and the space dimensions represented vertically. It shows two correlated particles emerging from a common source, moving off in opposite directions and moving forward in time. Local causality implies that the cause of the particles' presence and properties is at the source. Nonlocal causality implies the cause occurs elsewhere. Imagine a baseball in flight. Is the cause of a baseball's flight with the batter, or with the glove that catches the fly ball? Food for thought.

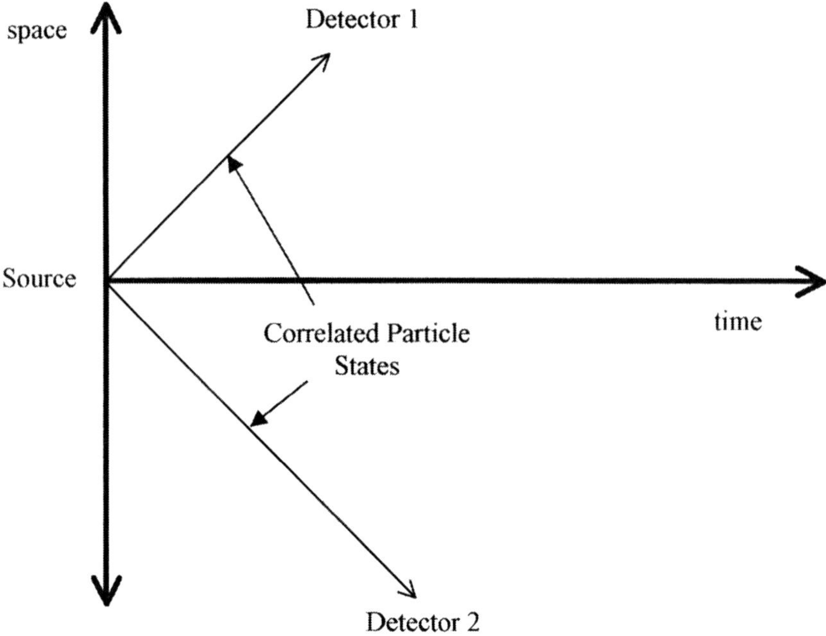

Figure 1. **Space–time diagram of two entangled particles departing from a common source and moving apart in opposite directions in space and forward in time.**

There are many demonstrations, or limited proofs, of Bell's theorem. The present treatment, adapted from Clauser, Horne, Shimony, and Holt (1969) (CHSH), is less abstract and more relevant to the experimental tests that have been performed. The CHSH abstractions have been mellowed by applying techniques used by Henry Stapp (1977)

Bell's theorem has been tested experimentally many times. Among the first were experiments performed by Freedman and Clauser at Berkeley (April 1972) and by Aspect, Grangier, and Roger in France (July 1982). Tests like these have shown that quantum mechanics rules and local causality fails.

We will give a limited demonstration of the theorem pertinent to an experiment with polarized photons. We show a nutshell version of Bell's theorem and its background. We show the derivation of the probabilities for quantum mechanical polarization states for two-photon emission, followed by a conceptual entangled particles experiment using two polarized photons emitted from an excited atom, much the same as was done in the original experimental tests. We show the quantum probabilities for measurement outcomes of the photon states, and discuss the local causality constraints on the outcomes. We review some of the early experimental evidence. Finally we give some philosophical implications of nonlocal causality. Among several hypotheses, that of causal time reversal emerges as the most cogent explanation for nonlocal causality.

A Few Definitions

We need to define some terms before we launch into the analysis.

• LOCAL CAUSALITY—This occurs when effect follows cause, staying within its light cone, and is contiguous to it in spacetime. This is causality as normally experienced, with which we are all familiar. It is a consequence of the assumption that elementary particles have an objective existence which persists throughout their spacetime paths.

• WAVE FUNCTION—(State function; State vector) A solution of the dynamical equations of quantum mechanics, describing the development in spacetime of the probability of a measurement outcome.

• SUPERPOSITION—A linear combination of wave functions, giving multiple possible measurement outcomes with different probabilities. Note that a superposition of states may have two interpretations. First is the existence of a single particle in a single undetermined dynamical state, the probability of which state is predicted by quantum mechanics. Or, it may imply that a distribution of potential states exists, only one of which is "born" when a measurement is made.

• WAVE FUNCTION COLLAPSE—A measurement process will result in one of multiple possible measurement outcomes in a superposition. The disappearance of the superposition, to be replaced by a single state, is called a collapse. The state is an eigenfunction, or a root function of the measurement.

• HIDDEN VARIABLES—Hypothetical dynamical variables of hidden particles, of which quantum mechanics gives only the probabilities. These are supposed to be real, but are not measurable with current technology.

• COPENHAGEN INTERPRETATION of quantum mechanics—The centerpiece of the philosophy of quantum mechanics, this interpretation says that there are no hidden variables. All measurements are probabilistic. An experimenter's choice of measurement will determine the form of the wave function as a solution of the dynamical equations. The interpretation is quite pragmatic. The Copenhagen interpretation was authored principally by Niels Bohr and Werner Heisenberg, and is so named because Bohr did his work at the Bohr Institute at the University of Copenhagen.

▪ ENTANGLED PARTICLES—These are widely separated particles in spacetime which belong to the same quantum state, or superposition of states, such that measurements on one of them are strongly correlated with measurements on the other. "Entanglement" is a term invented by Heisenberg to refer to the EPR paradox. Its use has been expanded to include other systems of the same genre.

▪ LORENTZ INVARIANCE—This refers to the invariance of the laws of physics under transformation from one relativistic uniformly moving reference frame to another.

The EPR Paradox and Bell's Theorem

Anyone who understands Bell's theorem
and isn't bothered by it has got rocks in his head.

Albert Einstein, the principle author of Einstein, Podolsky, and Rosen (EPR) (1935) considered the Copenhagen interpretation of quantum mechanics to be incomplete. To illustrate his case, he and his colleagues proposed a *gedankenexperiment.*

In this experiment, entangled particles with correlated states are emitted from a source, going in opposite directions. They impinge on two measuring detectors. The situation is represented in Figure 1. Until they are measured, the states are a superposition of several possible outcomes. When a measurement is made at Detector One, the state at Detector Two immediately collapses into the correlated state. This is in spite of the fact that the detectors are widely separated, and that there is no time for a light signal to pass between them. Einstein characterized this as "spooky action at a distance," and thought it to be impossible.

The EPR paradox is based on the assumption of local causality for the entangled particles. But Bell's theorem and the resulting mathematical inequality shows that quantum mechanical predictions of measurement outcomes for certain entangled states are inconsistent with local causality. We will show an example of how this may come about, which constitutes a limited proof of nonlocal causality. The discussion is an adaptation of Henry Stapp's treatment in *Il Nuovo Cimento* (1977).

Bell's theorem applies to the quantum mechanics of atomic particles. It was published almost fifty years ago in the first issue of *Physics*, and the journal folded after that one issue. For the next decade, the Bell paper seemed to the scientific world to be underwhelming.

The mathematical inequality has been tested repeatedly, and testing shows that quantum mechanics rules. Local causality often fails.

Quantum Mechanical Polarization States

The experimental concept to be described below depends on a linearly polarized light beam. Correlated photons are emitted from a common source, each polarized so that the electric field vibrates in the same plane. The emitted light beam, and each photon in it, has a superposition of all polarization states. If a measurement is made with a vertical polarizer, the superposition collapses into a single state, a vertically polarized beam.

A measurement of the new state at an off-angle does not, however, result in a null state. If a singly polarized state is passed through another polarizer at an angle theta, much like a force, the polarizer will capture its projection on the polarizer axis.

The analogy is that the wave function of the light beam is equivalent to an electromagnetic field, and an electromagnetic field is a force field. The polarizer is free to vibrate along the polarization axis under the influence of the force, but vibrations across the axis are forbidden by the structure of the polarizer molecules.

A Conceptual Entangled Particles Experiment

The experimental setup shown in Figure 2 was used by Stapp in his demonstration, and is also representative of the first experiments done to prove that quantum mechanical predictions prevail and locality fails.

The source is a vapor of calcium ions in a heated oven or an atomic beam, energetically emitting two correlated photons in opposite directions. The emission is a "cascade" process—one photon is followed quickly by the second. They share the same plane of polarization, one up and one down. The direction of polarization is an indeterminate superposition until it is measured. The polarizer axes are set at an angle of theta, one to the other. This angle is an important part of the subsequent analysis.

Local causality dictates that the source ion causes a photon to be emitted in a given direction and polarization state. This state is a "hidden variable." Local causality requires hidden variables. The photon incident on the polarizer is either absorbed or causes a new photon to emerge from the polarizer. The emergent photon, if there is one, causes a signal in the detector.

Each detector records an apparently random sequence of events. The current wisdom says that no information is transferred faster than light speed. The coincidences are not detectable until the two data records are brought together through classical information channels, after which they

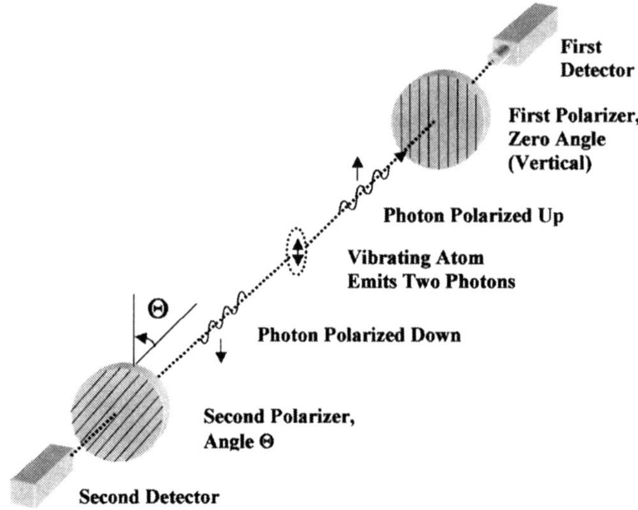

Figure 2.   **Experimental concept for coincidence measurement of entangled polarized photons.**

are compared. Then the correlations are apparent.

Recent experiments described in *Physics Today* (April 2011) indicate that information transfer may be "enhanced" with the use of entangled systems, which increase the efficiency of a transmitting channel to nearly theoretical values. The concept of "no information transfer" needs further review, and a careful definition of the meaning of "information."

Quantum Probabilities for Measurement Outcomes

Figure 3 is a template we use to display the probabilities, or normalized measurement rates, of two coincident photons through their respective polarizers. The circles with arrows in them are icons representing the polarizer orientations. The front and rear polarizer angles are shown to the left of these circles. The angle Θ is the included angle between the two polarization axes. The four large boxes each have a pair of polarizer settings representing the front and rear polarizers. The chosen angles were shown by CHSH (1969) to give the greatest violation of the Bell inequality.

Each of the four large boxes contains four small boxes, labeled **(a)** through **(d)**. Each of these in turn contains two circles, again representing the front and rear polarizers. They represent the four possible outcomes

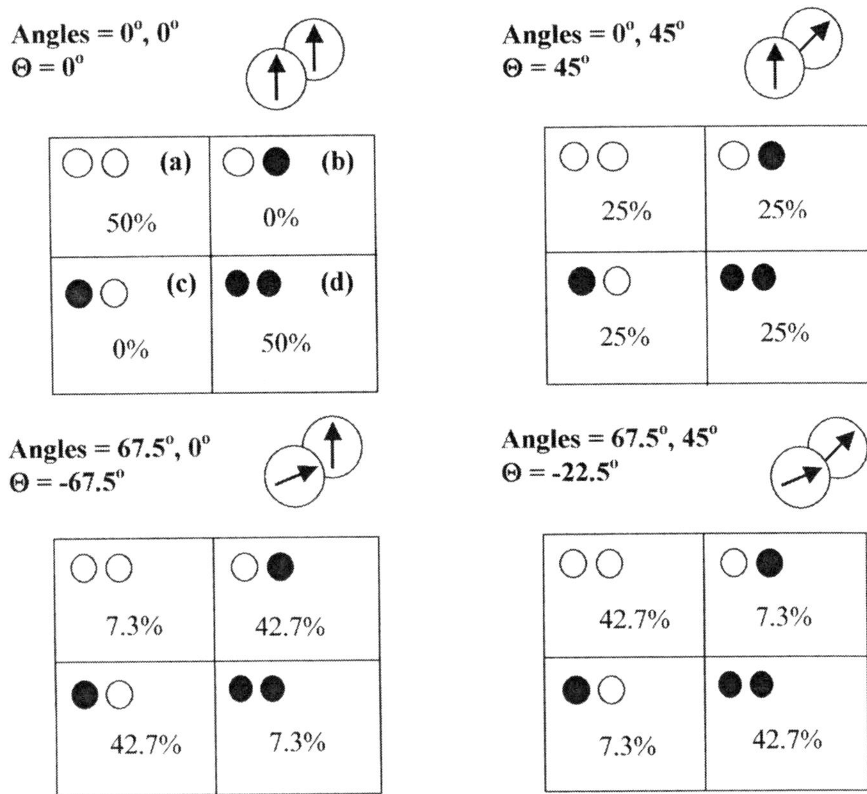

Figure 3. Template showing quantum mechanical measurement probabilities for four different polarizer settings and four different photon incidence event types.

of the photon measurements. The open circle represents transmission of a photon through the polarizer. The shaded circle represents absorption, or no transmission. Event **(a)** is a coincidence measurement. Event **(d)** is no measurement (null event). The other two are mixed transmission and absorption events which we will call anti-coincidences.

Figure 3 shows the measurement probabilities for each of the sixteen displayed states as predicted by quantum mechanics. The formulae used are $\frac{1}{2} \cos^2 \Theta$ in small boxes **(a)** and **(d)**, and $\frac{1}{2} \sin^2 \Theta$ in small boxes **(b)** and **(c)**. Notice that if the polarizers are parallel, every event appears either as a coincidence (open–open, 50%) or as an absorption of both photons (black–black, 50%). There are no anti-coincidence events.

The formulae may be derived using an incident quantum mechanical wave function, $e^{i\varphi}\sqrt{(2\pi)}$, with φ equal to the superposition of incident photon polarization angles. The expectation value of a complex coincidence operator, a function of the polarizer angle Θ, is involved.

A more intuitive approach may be taken. Suppose two incident photons succeed in passing through the polarizers. Then one or the other of the polarizers, with axis at an angle Θ, will have the photon's field strength reduced to its projection on the polarization axis. This gives a factor of $\cos \Theta$. This is a consequence of the force-like nature of the field strength, described in the above section on polarization states. Quantum mechanics tells us that the probability of seeing a measurement is proportional to the square of the field strength (or wave function) of the measurement. So the probability of seeing a coincidence in small box **(a)** is proportional to $\cos^2 \Theta$.

A careful analysis shows that the probability of a uniform superposition of polarizations passing a photon through a polarizer is one-half. That is, on average, half the photons are sufficiently aligned with the polarizer axis to get through and half are not. Consequently the proportionality factor of the coincidence measurement is ½, and the probability is $\frac{1}{2} \cos^2 \Theta$.

Suppose one polarizer transmits and the other absorbs. The probability of this anti-coincidence event together with a coincidence event is one-half. So the anti-coincidence must have a probability of $\frac{1}{2} \sin^2 \Theta$. This formula calculates probabilities for small boxes **(b)** and **(c)**. Conservation of probability then dictates the probability of the null event to be $\frac{1}{2} \cos^2 \Theta$. Then the sum of the four probabilities equals one for all polarizer angles.

Local Causality Constraints on the Outcomes

The squares of the trig functions derived in the last section will be used to calculate the quantum mechanical coincidence rates. These will be compared to local causality constraints, under the assumption that local causality is compatible with quantum mechanical predictions. Contradictions of these constraints with quantum mechanics will demonstrate that they are not compatible.

The procedure to follow will change each one of the polarizer settings sequentially to move from the upper left to the lower right polarizer settings (large boxes). As we do so, we examine the constraints of local causality for the changes in each coincidence rate (small boxes). The procedure is illustrated in Figure 4.

To facilitate the math, we need to define a fraction f, given in percent, of the total number of all measurements being moved along each of the two paths. This fraction would appear as the percentage of measured coincidences (open–open) at polarizer angle zero–zero which gets absorbed (black–black) as a result of changing the polarizer settings to 67.5 and 45

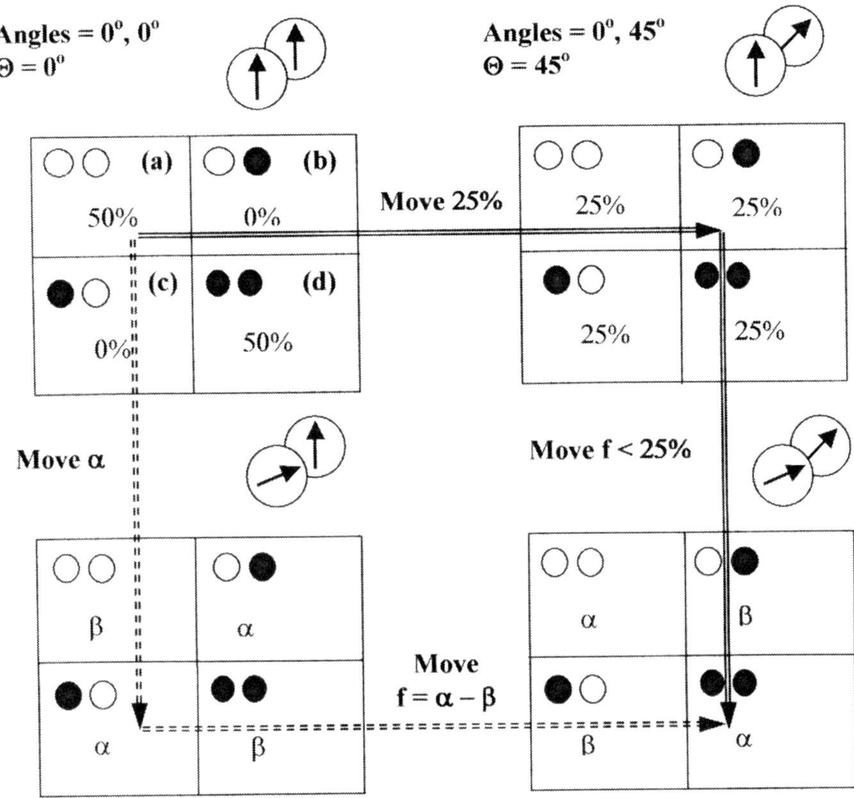

Figure 4.   **Template for analysis of local causality constraints; a fraction f moves from event type (a) to event type (d).**

degrees. The solid and dashed arrows represent two different sequences in which the polarizer angles are changed. The solid sequence changes the rear polarizer first, the dashed sequence changes the front polarizer first.

In Figure 4, follow the sequence of the solid arrows. To begin, only the second (rear) polarizer changes orientation. Imagine that for each polarizer setting the exact same set of photons emerges from the source and is incident on the polarizers. As we change the polarizer setting, we see how events are changed from coincidence events to anti-coincidence events. The key is that the locally caused photons are like real objects. Photons incident on the unchanged polarizer will transmit exactly as they did before. Photons incident on the changed polarizer will be partially blocked due to the polarization misalignment. Local causality allows measurement events in **(a)** to be converted only to events in **(b)**. Thus far, everything is assumed to be compatible with quantum mechanical coincidence rates. So the fraction of measurements moving from coincidence to anti-coincidence, by the ½ $\sin^2 \Theta$ formula, is 25%.

Next, change the front polarizer to a new angle of 67.5 degrees. Then local causality will allow changes only in the transmission rates in the front polarizer, and measurements can move only from small box **(b)** to small box **(d)**. The fraction of measurements f moving from coincidence to null (pass–pass to block–block) is 25% at most.

Now we are going to have locality constraints on the incidence rates appearing in the two lower large boxes, and the results are unknown. Accordingly, we change the numbers to algebraic symbols. But we still assume compatibility with quantum mechanics. Compatibility requires that α and β trade places between the lower two large boxes in Figure 4. Conservation of probability also requires that α and β add up to 50%. Following the dashed arrows, we first change the front polarizer to an angle of 67.5 degrees. Due to local causality, the same photons will still be passed by the rear polarizer, but some of the coincident photons on the front polarizer will now be blocked.

Local causality allows the coincidence measurement events in **(a)** to only be converted to anti-coincidences in **(c)**. All of the α events now in **(c)** must come from **(a)**. Next, change the rear polarizer to a new angle of 45 degrees. Then local causality will allow changes only in the number of events detected in the rear polarizer, and measurements can move only from small box **(c)** to small box **(d)**. The fraction of measurements moving is $\alpha - \beta$, since β of them must remain in small box **(c)**.

Note that the solid and dashed path protocols may be interchanged, obtaining the same results.

Two of the constraints on f are as shown in Figure 4. The third constraint on $\alpha + \beta$ derives from the symmetry of α and β and conservation of probability, that all incidence rates must add up to 100 percent. In summary, the constraints are:

$$f < 25 \qquad\qquad f = \alpha - \beta \qquad\qquad \alpha + \beta = 50$$

Eliminating f from the three equations gives a solution for α and β.

$$\alpha < 38 \qquad\qquad \beta > 12$$

This appears as an inequality constraint, and is a special case of the Bell inequality. These two constraints now apply in the bottom two large boxes of Figure 4, where the front polarizer orientation is at 67.5 degrees.

Comparing these with the quantum mechanical predictions for the same polarizer settings shown in Figure 3, it is apparent that the quantum mechanical predictions violate the local causality constraints.

This demonstrates the incompatibility of local causality with quantum mechanical predictions for entangled, polarized photons, under the particular polarizer conditions chosen.

We may ask why we can't write an exact equality for the fraction f rather than the inequality we have shown above. This would be a strong constraint on the compatibility of quantum mechanics with local causality, rather than the weak (inequality) constraint shown. We could do this. We have in fact imposed a strong constraint by demanding in Figure 4 that the measurement probabilities for quantum mechanics and local causality be equal for parallel polarizer axes (upper left large box). But local causality could be compatible with quantum mechanics even within a range of measurement probabilities. So we have weakened the constraint in passing to the large box on the right. We did this with the intention of solving for the range of compatibility.

Experimental Evidence for Nonlocality

Many experimental tests have been done to validate quantum mechanics and to demonstrate that entanglement leads to nonlocality of cause and effect.

These tests give rise to considerable tension between two sacrosanct theories, quantum mechanics and local causality. These are two of the finest of fundamental physics. They are at odds with each other and will not coexist. Which will the test prove to be correct? From the beginning, there was really no question. Nobody beats quantum mechanics.

The landmark tests were done with cascaded polarized photons from ionized calcium vapor. The first test, the breakthrough, was the Clauser–Freedman experiment (1972) done at Berkeley.

There were a number of criticisms of the experimental controls, the lack of which might have provided an (unlikely) opportunity for misinterpretation. The results were statistically significant.

Aspect performed what is now considered the defining test, with tighter experimental controls, at the University of Paris in Orsay. His experiment included changing the polarizer settings while the photons were on the fly. With passing years, the tendency is to refer only to the Aspect experiment and, perhaps unfairly, to overlook the original Clauser–Freedman experiment.

Philosophy of Nonlocal Causality

Are we not drawn onward, we few, drawn onward to a new era?

If the cause-and-effect does not begin at the source, with the emitting ions, where is it? Is the photon caused by the observation? Is the cause at the detector? This explanation is problematic. But perhaps we are misled, as we are by the palindrome, into thinking time has only one direction. However, let us first consider a few of the more popular alternatives.

A common explanation, perhaps the most widely used, is that entangled particles form a single quantum state, and will respond to a measurement as any state does, by yielding a sub-state out of the superposition. This is pure, pragmatic Copenhagen interpretation, and really begs the question of the structure of cause and effect.

One might argue that causality as a physics principle is outside the realm of spacetime. Then its true structure remains to be discovered, as does its mode of interaction with spacetime. Another argument holds that causality is a human-derived artifact, an appearance of a larger reality, useful only in describing our experience of spacetime.

On the other hand, if there are hidden causal interactions between the two detectors and the source, such as David Bohm's pilot wave, their responses may be guided as specified by quantum mechanics.

In a variant of his experiment, Alain Aspect (Aspect, Dalibard, & Roger December 1982) modified his apparatus so that the polarization of Detector One was set after the photons had left the source. The polarization change was accomplished in a pseudo-random fashion, so it is difficult to imagine how hidden causal interactions could govern the final detector dynamics before the source emitted its photons.

The tentative conclusion from the hidden causal interactions hypothesis is that if there is a hidden transmission of causal interaction from one detector to the other it must be "superluminal" (faster than light speed). In Figure 1, this would correspond to an arrow pointing from one detector to the other. But there is one other possibility.

Figure 1 shows arrows pointing away from the source, indicating a timewise forward motion of the particles, and implying a causal link forward in time. If one of the arrows pointed back toward the source, this could be taken as a causal link starting at the detector, which more or less randomly collapses the wave function into a single polarization state, which then collapses the source into that state in a retrocausal fashion. The unique state thus assumed by the source will then cause the measurement at the other detector to be suitably correlated. This is time reversal of cause and effect.

The symmetry of dynamic time reversal is well-known in field theory. There are no theoretical reasons that would rule out causal time reversal if dynamic time reversal is possible.

Causal time reversal is attractive because it maintains contiguity through the photonic paths, and the process remains Lorentz-invariant, requiring only that we use improper Lorentz transformations where the time element of the metric is negative instead of positive.

The problem is, causal time reversal is anathema to many physicists. It is way outside the paradigm, it seems, and many cannot grasp the concept. The widespread concept is that cause precedes effect by simple definition. Our life experiences have left us with neural patterns that preclude the opposite concept.

If we accept causal time reversal, we must ask why the primary cause should be at Detector One, and not at the more distant Detector Two. Perhaps both detectors contribute to causality in ways dependent on the experimental/observational setup.

No one has thought of a good objective physical test that will prove or disprove causal time reversal. But causal time reversal may emerge on the macroscopic level, especially when certain types of cognitive action comes into play. Causal time reversal may explain some types of cognitive action yielding foresight or premonitory effects. Dean Radin (1997) has done experiments to show presentiments of startling events using galvanic skin reaction measurements. Dobyns (2006, AAAS San Diego meeting) has shown that certain types of premonitions or precognitive processes require causal time reversal to explain them. Some other types of rare processes involving apparent entropy decrease in evolving systems also seem to fit a causal time reversal paradigm.

In retrospect, nonlocal events may nearly always be accompanied by time reversal of cause and effect. They are, in a way, synonymous with each other.

References

Aspect, A., Dalibard, J., & Roger, G. (December 1982). Experimental test of Bell's inequalities using time-varying analyzers. *Physical Review Letters, 49,* 25.

Aspect, A., Grangier, P., & Roger, G. (July 1982). Experimental realization of Einstein–Podolsky–Rosen–Bohm *Gedankenexperiment*: A new violation of Bell's inequalities. *Physical Review Letters, 49,* 2.

Bell, J. S. (1964). On the Einstein–Podolsky–Rosen paradox. *Physics, 1,* 3, 195–200.

Clauser, J. F., Horne, M. A., Shimony, A., & Holt, R. A. (1969). Proposed experiment to test local hidden-variable theories. *Physical Review Letters, 23,* October, 15.

Dobyns, Y. (2006). Retrocausal Information Flow: What Are the Implications of Knowing the Future? In D. Sheehan, Editor, *Frontiers of Time: Retrocausation—Experiment and Theory,* New York: AIP Press. In *Proceedings of the 87th Annual Meeting of the Pacific Division of the American Association for the Advancement of Science*; San Diego, CA; June 18–22, 2006.

Einstein, A., Podolsky, B., & Rosen, N. (1935). Can quantum-mechanical description of physical reality be considered complete? *Physical Review, 47,* 777–780.

Freedman, S. J., & Clauser, J. F. (1972). Experimental test of local hidden-variable theories. *Physical Review Letters, 28,* 1.

Radin, D. I. (1997). Unconscious perception of future emotions: An experiment in presentiment. *Journal of Scientific Exploration, 11*, 2.
Stapp, H. P. (1977). Are superluminal connections necessary? *Il Nuovo Cimento, 40*, 1.

Magnetism and the Brain

Take as an example the ultimate claim, that individual brain functions may be controlled by suitable waveforms of magnetic field applied to appropriate places on the skull. For this to be taken seriously, it is reasonable to expect at least some suggestion of a feasible explanation as to how this may be achieved. But in the current published literature on parapsychology, while there are frequent references to discoveries by neuropsychologists of the location of particular brain functions, there is an almost complete absence of any reference to discoveries by neurophysiologists on the structure or function of the stream of pulses ("action potentials") that carry the "signal" in the axon of a brain cell—which is the process claimed to be controlled.

Consider the facts of the structure of the neocortex of the brain. It consists of a layer of neural material about 3 mm thick, each square millimeter of surface covering about 200,000 neurons ("nerve cells"). If the magnetic field from an external electromagnet affects one square centimeter of cortex surface, this includes 2×10^7 (twenty million) cells. These are organized in "columns" of associated neurons about 0.3 mm diameter, so about 1,000 columns are affected, each of which (in principle) can be carrying out a different function (Szentágothai 1989). To apply a magnetic field to one neuron, or a small group, or even a specific column, without affecting other functions, would seem an impossible task.

The "axon" (nerve) of each cell (*whatever its function*) carries a stream of *identical* quasi-electrical triangular pulses of about 1 ms (millisecond, one thousandth of a second) duration (Szentágothai 1989, Ralphs 1995), and it is the "pattern" of bursts of pulses (or, more probably, in the author's opinion, the erratic gaps between bursts) that presumably "carries the message." So, assuming that it was possible to inject an artificial signal into a single axon, what form should it take?

Low-frequency magnetism has virtually no effect on the non-magnetic materials that constitute the human body. For instance, the very strong fields used in MRI scanning have little or no effect. Higher frequency stimulation (such as attempts to duplicate natural neural waveforms) would require very accurate, rapidly changing waveforms to be injected into specific minute structures within the cortex.

At best it may be possible to use strong fields in a purely negative sledgehammer process (just as one can stop a mechanical watch escapement, the spark-ignition system of a car engine, or the conversation on a telephone line by the same means) to disable or corrupt several neighboring functions. As an example, if you could disable the functions of fear, doubt, and worry in the brain, the result would be a feeling of utter peace and security—a true nirvana that, with a little gentle suggestion, could easily be interpreted as an awareness of the presence of angels, or even God himself. This could hardly be called "control," although it is notable that many such claims can be explained in similar negative terms. With such facts in mind, it is evident that attempts to control neural systems by magnetic fields are misplaced, to say the least.

Another error frequently committed is to regard biologically produced electrical fluctuations (such as EEG traces) as potential "electromagnetic (EM) waves." A fluctuating electric current or magnetic field does not immediately produce an electromagnetic wave. At a distance from its source, the electrical and magnetic energies in a true electromagnetic wave are equal, but close to the source one or the other is predominant (depending on the type of source), and the wave must travel for more than half a wavelength before the two energies are within 20% of each other (ITT 1956). Accepting this as a "definition" of a true EM wave, if a wave of frequency "f" KHz is at a distance "s" Km from its source, and f × s is less than 150, it is not yet a true EM wave, merely fluctuating magnetic and/ or electrical fields. When you consider that the significant frequencies in an EEG trace are well below 30 Hz (which has a wavelength of 1,000 Km), it evidently fails to qualify as an EM wave at any reasonable range.

Magnetic Storms

The theory that fluctuations in the local magnetic field can cause hallucinations or other mental phenomena seems to have originated in published articles that identified correlations between periods of strong magnetic disturbances and the reporting of hallucinatory visions (Randall & Randall 1991), and poltergeist and other PK phenomena (Gearhart & Persinger 1986), which apparently confirmed that magnetism can affect neural functions. However, it is very probable (and in my opinion almost certain) that this assumption is not so much incorrect as seriously misleading. My reasoning is based on the structure and origin of so-called "magnetic storms" (a subject that seems to have generated a mythology of its own).

It has been known for more than 150 years that the appearance of "spots" on the surface of the sun could be accompanied by inexplicable deviations of a ship's compass, and as a result the strength and direction of the earth's magnetic field is studied internationally and has been continuously monitored and recorded in the UK since 1868, the data being collated nowadays by the British Geological Society. Apart from its interest to scientists, this information is gathered for very practical reasons. Such magnetic surges can induce electric currents of considerable strength into the electrical mains supply network. This would be immaterial, except that control signals between major power stations are passed over the distribution network itself and if these signals are blocked or corrupted, control can be lost or misapplied, causing major failures of electricity supplies. These disturbances follow an eleven-year "sunspot cycle," and those in America during the last sunspot maximum blacked out half the USA (Beamish, Clark, Clarke, & Thomson 2002). The next maximum is due at about the end of 2012.

In general, magnetism originates from a permanent magnet or an electrical device generating a magnetic field. In either case the source has two opposing "poles." "Lines of Magnetic Force" leave the source from its North pole and take the shortest route open to them (subject to certain "laws") to re-enter the source at the South pole. This means that it is essentially a short-range phenomenon, limited to five or six times the distance between the two poles. This applies to magnetic storms on the sun, just as much as to a pocket magnet. Therefore any magnetic field on the surface of the sun will not reach the earth, and another mechanism to explain the observed correlations is necessary.

The "mechanics" of such storms were established by the study of "cosmic rays" in the early twentieth century (Millikan 1939). They were found not to be "rays" in the normal sense of the word, but dense streams of electrically charged particles—bits of smashed-up nuclei of atoms—but the title is still used. They are continuously emitted by the sun (and other sources in space), but in massive quantities from a sunspot, which can be imagined in terms of a gigantic volcanic eruption ejecting millions of tons of this electrically charged "dust" at speeds in excess of four million miles per hour [sic!] (a Coronal Mass Ejection (CME)). As they approach the earth, these particles are deflected by the earth's magnetic field toward the poles, creating the beautiful Northern Lights (Aurora Borealis), but this deflection is not complete, and more than 70% of the stream reaches earth, even at the magnetic equator. [Ryan's description of "an electrically charged gas" (Ryan 2008) is extremely misleading, but understandable, since the term *plasma* is often applied to a cloud of electrically charged particles, and some dictionaries define the word as "a hot gas." In this case the difference is vital.]

The effects of these particles on earth must be considered with reference to their ability to penetrate deeply into material objects. Every human head in the world has more than a hundred of them travelling right through it every minute, and the more powerful particles can go through several centimeters of lead or about ten meters of sea water. Each particle is so small that only one in several million will actually collide with the nucleus of an atom, but when it does, it can split it into two or three parts (Millikan 1939:53ff). A particle can "zap" a transistor (being the only known "wear-out" mechanism for the computers in a spaceship and so of great interest to NASA). So one would expect it to be quite capable of affecting a cell in the human body or brain, either by physically damaging it, or by passing so close to an axon as to set up a large (but extremely brief) magnetic pulse in it.

Alternative Explanations

When an electrical charge moves, it generates a ring of magnetic force around its path with a strength depending on the strength of the charge and the speed at which it is moving. The quantity and velocity of cosmic rays means that they can generate quite appreciable magnetic fields at the surface of the earth. So the two phenomena, the electrical charges and the magnetic field they generate, are correlated. It is not surprising that a correlation exists between severe magnetic activity and some mental factors (including ESP), but *it is a distinct possibility—indeed, a definite probability—that the active agent in most such cases is NOT the magnetic fluctuations themselves, but the cosmic rays that cause them*. For instance, the electricity supply problem is attributable to the former, while the considerable effects on the earth's ionosphere are predominately due to the latter. In the days of Short-Wave (HF) signalling, an SID (Sudden Ionospheric Disturbance) could disrupt long-range communication for a period from half an hour to half a day by particles destroying the ionospheric layers over a very wide area.

Magnetism is relatively easy to generate, to control, and to measure; while charged particles are far less amenable. Inevitably in virtually all studies (in all fields) it is customary to refer to magnetic measurements and to ignore or forget the particle aspect. The fact that there is a close correlation between the two, the particles generating the magnetism, makes this unimportant in most cases, but in the case at issue the distinction is vital. All laboratory investigations on the lines indicated by Persinger and others (Persinger, Tiller, & Koren 2000) apply magnetic fields alone to the brain, so only one of the active agents is being replicated. If, as I suggest, the charged particles in a magnetic storm may be equally or even more effective, any results, positive or negative, could be seriously misleading.

Physiological Correlations

Because of the regularity of the eleven-year sunspot cycle and the existence of detailed medical records over many years, it is relatively easy to trace correlations between magnetic storms and a number of physiological variables, including:

Positive Correlation: Suicides, depression and mental disorders, heart rate and high blood pressure, SIDS (Sudden Infant Death Syndrome).

Negative Correlation: Melatonin secretion in the pineal gland, circadian rhythm, sensitivity to light, fatal heart attacks, etc. (Ward & Henshaw).

It follows that, while correlation between magnetic anomalies and various aspects of ESP is a reasonable assumption, it is far from exclusive, and the wide range of effects suggests that the basic mechanism is almost certainly mechanical or electrical at the cell level, with no explicit paranormal associations. It follows that theories postulating precisely defined complex "magic wiggles" of magnetism generating hallucinations by performing specific functions in the brain (Braithwaite 2004, Persinger, Tiller, & Koren 2000) are not only unjustified and unsupported, but may be considered as bordering on the tendentious.

EEG and MEG

A second serious criticism of many experiments is the total reliance on EEG and MEG (magnetoencephalogram) recordings. The fact that such waveforms do not represent any functional neural activity anywhere in the brain is generally ignored.

The basic "action pulse" which is the carrier of neural information throughout the whole of the human nervous system is a symmetrical triangular ("isosceles") pulse with a base width of about 1 ms, and each is followed by a chemical "dead time" of about 1 ms, during which another pulse cannot be generated. So in theory the fastest pulse stream which can be generated has a period of about 2 ms, at a frequency of about 500 Hz. The information seems to be carried by short and long bursts of pulses at something like this frequency. (The signals recorded from individual nerves in a rabbit (Barlow 1987) show some single pulses, but many bursts of 20 to 100 successive pulses at a frequency of about 100 Hz.) So one would expect a Fourier analysis of the signal in a single human neuron to show a broad maximum somewhere below 500 Hz (Appendix 1).

In contrast, EEG recordings show no pulses, but an irregularly varying voltage at much lower frequencies, with nothing significant above 30 Hz. They are generated by a "clearing-up" process in the brain. Transfer of electrical pulses along nerves inevitably affects the distribution of electric charges in the brain, and the resulting imbalances are corrected by normal electric currents flowing back through any available return path, such as blood vessels or bone. This includes the bone structure of the skull, which allows them to be detected by EEG electrodes. Their characteristics vary according to the ***level of neural activity*** in the area of the electrodes, which makes them invaluable to neurophysiologists ***in diagnostic and "brain-mapping" work***, but if the stimulus is unknown there is no information in the waveform itself on the actual neural process being carried out or its significance.

This is particularly true of the MEG, as a study of the generation of an action potential pulse establishes that in theory it creates no magnetic field at all. The extremely low level of field detected in practice supports the view that this is created by fortuitous epiphenomena (secondary divergences from the mathematical model), which are unlikely to carry much more useful information than the EEG. It should also be noted that normal Faraday screening is ineffective against low-frequency magnetic fields, as correctly noted by Ryan (2008). He also notes that:

> ELF [Extremely Low Frequency] spherics (the standing waves surrounding the earth, continuously powered by lightning strikes) in the 5–50 Hz frequency range are known to be disrupted by GMA [Geomagnetic activity]. (Ryan 2008)

This author would question this concept. Two magnetic fields from independent sources can add only vectorially and arithmetically. Lightning static is not affected appreciably by cosmic rays, or vice versa, and is virtually continuous at all times all over the earth. It acquires characteristic changes to its frequency structure by repeated travel around the circumference of the globe (allied to the so-called Schumann Resonances). Major solar flares can generate considerably higher levels of magnetism, but with totally different characteristics. It may make the smaller signal more difficult to observe, but this is irrelevant.

It is extremely rare for this distinction to be mentioned in parapsychological literature. Braithwaite (2004) for instance recorded signals with all the known characteristics of lightning static, but still (wrongly) attributed them to possible solar hallucination-generating anomalies. It is a reasonable conclusion that he is not alone, and many analyses by parapsychologists of suspected magnetic disturbances fail to distinguish between the two (which, since lightning static is not correlated to sun-spot activity or vice versa, could throw doubt on some conclusions from the correlation studies mentioned above).

Conclusions

This article is intended to be helpful and informative, but by implication the underlying state of affairs described is not encouraging. It suggests that the academic study of parapsychology is being carried out in an introspective and self-satisfied manner by specialists who cannot or will not recognize that, whether they like it or not, study of the paranormal by modern means must eventually involve sophisticated electronic equipment carrying out physical measurements of physical quantities that obey physical laws. B. F. Skinner and computerized statistics cannot provide all the answers.

Note

[1] Note that I am not suggesting that there is no correlation between magnetic fields and the paranormal (indeed, I consider it quite possible), but I doubt the ability of present-day techniques in academic parapsychology to investigate the matter effectively.

References

Barlow, H. (1987). *The Biological Role of Consciousness.* In C. Blakemore & S. Greenfield (Editors), *Mindwaves,* Oxford: Basil Blackwell, p. 362.

Beamish, D., Clark, T. D. G., Clarke, E., & Thomson, A. W. P. (2002). Geomagnetically induced currents in the UK; geomagnetic variations and surface electric fields. *Journal of Atmospheric & Solar-Terrestial Physics, 64,* 1779–1792.

Braithwaite, J. J. (2004). Variances associated with "haunt-type" experiences: A comparison using time-synchronised baseline measurements. *European Journal of Parapsychology, 19,* 3–28.

Gearhart, L., & Persinger, M. (1986). Onset of historical poltergeist episodes occurred with sudden increases of geomagnetic activity. *Perceptual & Motor Skills, 62,* 463–466.

Hinton, C. H. (1904). *The Fourth Dimension.* London: George Allen & Unwin.

ITT (1956). *Reference Data for Radio Engineers.* New York: International Telephone & Telegraph Corp. p. 663.

Millikan, R. A. (1939). *Cosmic Rays.* Cambridge University Press.

Persinger, M. A., Tiller, S. G., & Koren, S. A. (2000). Experimental stimulation of a haunt experience and elicitation of paroxysmal electroencephalographic activity by transcerebral complex magnetic fields: Induction of a synthetic "ghost"? *Perceptual & Motor Skills, 90*(2), 659–674.

Ralphs, J. D. (1992). *Exploring the Fourth Dimension.* UK: Foulsham, 0-572-01624-7. US: Llewellyn.

Ralphs, J. D. (1995). Assessment of factors affecting the detectability of thermal radiations from the neural system of the brain. *Medical & Biological Engineering & Computing, 33*(3), 264–270.

Randall, W., & Randall, S. (1991). Solar wind and hallucinations. *Bioelectromagnetics, 12,* 67.

Ryan, A. (2008). New insights into the links between ESP and geomagnetic activity. *Journal of Scientific Exploration, 22,* 335–358.

Szentágothai, J. (1989). The "Brain–Mind" Relation: A Pseudo-Problem? In C. Blakemore & S. Greenfield (Editors), *Mindwaves,* Oxford: Basil Blackwell, pp. 323–336.

Ward, J. P., & Henshaw, D. L. *Geomagnetic Fields, Their Fluctuations and Health Effects.* Jonathan. ward@Bristol.ac.uk

Zöllner, J. C. F. (1880). *Transcendental Physics.* London: W. H. Harrison.

Appendix 1
Fourier Analysis of Brain Axon Activity

Consider a stream of symmetrical triangular pulses of amplitude A and base width = 2t, repeated at time intervals T. The Fourier Analysis of this pulse stream (ITT 1956:1019 Figure 6b) gives a stream of sinusoidal components of amplitude Cn.

$$Cn = 2Av \left[(Sin\ x) / x \right]^2$$

where: Av = Average Value, $x = (nt/T)\pi$, and n is any integer.

Now consider the repetition time T to be freely variable.

The smaller T, the higher the pulse repetition frequency, so more pulses per second and the larger the Average Value.

If (nt/2T) is an even integer, Sin x = 0 and Cn = 0.

If (nt/2T) is an odd integer, Sin x = 1, its maximum value, and Cn is then proportional to $(1/n)^2$.

So one would expect the largest component to occur when n = 1 and T = 2t, but, in fact, this cannot happen. The pulse is about 1 ms duration at base, so t = 0.5 ms and there is a chemical "dead period," slightly longer than a millisec, after the pulse, in which it is difficult (impossible?) to trigger a second pulse (Szentágothai 1989). So T cannot be less than about 2 ms or 4t. So I would expect a strong component at about n = 5, making T about 2.5 ms and frequency about 400 Hz, modulated by quasi-random variations with roughly a Sin distribution. In fact, recordings (in animals) of single nerve patterns often show bursts of pulses at about the maximum frequency, so the noise component may often be lower. I have no knowledge as to whether such measurements have been carried out (or are possible?), and have been out of touch with the field for more than twenty years.

As discussed in this article, the acceptance of EEG traces as representing neural signals is completely misleading, since they are the result of a "clearing up" process which is restoring the electrical balance in the brain by returning electrical charges that have been moved in the normal neural processes. No EEG trace bears the slightest resemblance to any signal in a neural axon.

Journal of Scientific Exploration, Vol. 26, No. 4, pp. 791–824, 2012   0892-3310/12

RESEARCH ARTICLE

NDE Implications from a Group of Spontaneous Long-Distance Veridical OBEs

ANDREW PAQUETTE

paqart@gmail.com

Submitted 2/9/2012, Accepted 7/26/2012

Abstract—The case for veridical out-of-body experiences (OBEs) reported in near-death experiences might be strengthened by accounts of well-documented veridical OBEs not occurring near death. However, such accounts are not easily found in the literature, particularly accounts involving events seen at great distances from the percipient. In this article, I seek to mitigate this paucity of literature using my collection of dream journal OBE cases. Out of 3,395 records contained in the database as of June 15, 2012, 226 had demonstrated veridicality. This group divides into examples of precognition, after-death communications, and OBEs. Of the OBEs, 92 are veridical. The documentation involved is stronger than is normally encountered in spontaneous cases, because it is made prior to confirmation attempts, all confirmations are contemporaneous, and the number of verified records is large relative to the total number of similar cases in the literature. This database shows that NDE-related veridical OBEs share important characteristics of veridical OBEs that are not part of an NDE. Because the OBEs are similar, but the conditions are not, skeptical arguments that depend on specific physical characteristics of the NDE—such as the use of drugs and extreme physical distress—are weakened. Other arguments against purported psi elements found in veridical OBEs are substantially weakened by the cases presented in this article.

Keywords: near-death experience—out-of-body experience—veridical— OBE case history—dreams—psi

Introduction

Reports of near-death experiences (NDEs) have excited a great deal of popular and scientific interest in the years since 1975 when Raymond Moody first coined the term in his book *Life after Life* (Moody 2001). These experiences involve a group of unusual phenomena that typically occur in moments of crisis, particularly when there is serious threat to life. Some researchers and many NDE experiencers claim that the elements of a typical NDE appear to be a spiritual experience rather than a purely physical one (Bonenfant 2000, Dell'Olio 2010). Those who hold the dualist perspective believe that NDE data indicate a nonphysical element to consciousness, whereas those holding the nondualist perspective believe the data are explained by physical factors alone (Lundahl 2000).

One of the five major features of an NDE is "a sense of detachment from the physical body or sensation of floating out of the body, in which experiencers may find themselves looking down on their physical body and surroundings" (Schwaninger, Eisenberg, et al. 2002:215). This feature is referred to as an out-of-body experience (OBE). Because of the mental impression of displacement from the actual location of one's body, some skeptics of the dualist explanation for NDEs have developed the theory of dissociation, wherein a patient creates a mental image of one's body as viewed from an external position (Greyson 2000).

For dissociation to be valid, the subject must be able to construct a mental image. Patients who are clinically dead due to prolonged absence of brain activity and cessation of heart function should not be able to do this if consciousness is wholly dependent on the brain (Lommel, Wees, et al. 2001). Critics agree, but dispute whether NDEs actually occur during the period when the patient has a flat EEG reading (French 2001). This argument was used to explain the case of Pam Reynolds, who underwent radical surgery that temporarily left her, in every way measurable by modern medical science, dead. To prepare for the surgery, Reynolds' blood was chilled in a cardiopulmonary bypass machine, and then returned to her body. After 20 minutes, her core temperature dropped to 60 degrees Fahrenheit. This was done to intentionally stop her heart and brain activity. Once this objective was reached, the blood from Reynolds' body was drained (Sabom 1998). Despite this, critic Keith Augustine has complained that Reynolds' veridical OBE occurred prior to the pulmonary bypass, at a time when she was under general anesthesia, and may have been capable of receiving sensory information, despite heavy sedation (Holden 2009). This, he asserted, is enough to justify a physiological explanation for Reynolds' veridical perception.

Background

Veridical OBEs can be used to counter objections from skeptics, who agree that they represent a potentially strong challenge to physical explanations of NDEs (French 2001:2010–2011). To make a case for genuine veridicality, skeptics have asked for evidence that environmental ambience is not a factor. To answer this, studies of blind patients who experienced veridical NDEs have been conducted. With these studies, researchers sought to show that people who have had suboptimal visual sensitivity could accurately describe visual attributes of objects that they are physically incapable of seeing (Irwin 1987, Ring & Cooper 1997, Long 2010). These should be enough to quell criticism, but when reports of the cases are examined carefully, researchers discover that there are not as many cases as they expected to find, and the cases are not well-supported by contemporaneous documentation (Ring & Cooper 1997). Without contemporaneous documentation, they fall prey to criticism of memory errors among witnesses.

In a literature review that attempted to locate every example of a verdical, NDE-related OBE contained in the scientific literature, a total of 107 cases were found, the oldest of which occurred in 1882 (Holden 2009). Of this group, only 17 included an interview of the subject within two days of the NDE. The lack of contemporaneous documentation in the remaining 90 cases provides opportunity for skeptics to complain about weak documentation. Laboratory-conducted OBE studies feature accurate real-time records, but they are rare (Tart 1998).

The problem presented by most veridical OBE cases in the literature is that they do not describe a long-distance OBE *and* have contemporaneous records of the event. Without both of these elements, the cases are susceptible to criticism that some unknown sensory capability of the body has produced the apparently veridical OBE in a normal, purely physical manner or that they are simply the result of some kind of memory error that is impossible to check due to the lack of records.

Personal OBE Database and Bias

Spontaneous cases of veridical OBEs do occur, but when they make their way into the literature, they normally are not accompanied by corroborative contemporaneous documentation. With this paper, I would like to introduce some exceptional cases of well-documented, long-distance, spontaneous OBEs. Each of these examples is a personal experience of mine that was recorded in writing prior to verification of veridical elements. It is my hope that by introducing these cases of non-NDE OBEs, veridical aspects of the near-death OBE can be better understood.

The collection of cases officially began on September 16, 1989. Like the engineer and author J. W. Dunne (Dunne 1927), I began by keeping a dream journal. Unlike Dunne, I did not do so because I had noticed that my dreams seemed to be precognitive but rather because my wife had noticed such a pattern and I didn't believe her. I will admit, however, that the idea was as tantalizing as it seemed ridiculous. Regardless, without any hope of success, I began the journal. Any bias I may have had when the journal started was antagonistic to a psi explanation. At the time, I assumed that poor memory for earlier unrecorded dreams prompted my wife to erroneously conclude that I had experienced precognitive dreams.

The Journals

For the first 365 days the journal was kept, the period of September 16, 1989, to September 15, 1990, 375 records were made. The number of records exceeds the number of days in the year because of days when separate records were made for overnight dreams and daytime naps. Of the 375 records, 114 had veridical content. Of the 114 records that contained veridical items, 64 were OBEs, of which. 50 were veridical. The number of veridical items drops sharply after the first year because I stopped actively checking records for veridicality after becoming satisfied that the journals had captured genuine psi content. If verification of dream items was given to me, I would usually record it, but passive verification led to a smaller number of verifications. Despite this, 226 of the first 3,395 records are veridical. For the full period covered in the first 32 journals, September 16, 1989, through June 15, 2012, there were 8,311 days. The 3,395 records cover 41% of that period. Table 1 provides a list of the number of records containing veridical items per journal, along with the number of records contained in each journal.

The percentage of veridical records per journal drops sharply after Journal #3 (DJ03). The cause of this change is that I stopped actively checking dreams for veridicality after DJ03 was complete.

Table 2 provides a breakdown of veridical psi events recorded over the first year in the journals. The same group of records contains probable psi material that either could not be verified, or verification was not attempted. Some of these are after-death communications, purported past-life memories, lucid dreams involving spiritual instruction, and remote psychokinetic healing. None of these are included in Table 1, because those records were not validated. OBEs that involve classic signs of a veridical OBE but that do not contain sufficiently unusual information are also not counted as veridical. The most common of these are lucid dreams where I see my sleeping body in bed. Because I was familiar with the contents of my apartment at the time, no amount of accuracy or detail could distinguish such an OBE from normally acquired knowledge.

In Table 2, veridical dream records from the first 365 days of the journal are coded based on whether they are an OBE, precognitive, or something else. Examples of "something else" are two records from July 4 and July 5, 1990, that had veridical content related to reincarnation. Three records from Journal 3 had examples of two different types of psi in each. This is why the number of records containing veridical psi is 84, but the tally in Table 2 is 80.

TABLE 1
Incidence of Veridicality per Journal

Dream Journal ID	Number Veridical	Number of Records	Dream Journal ID	Number Veridical	Number of Records	Dream Journal ID	Number Veridical	Number of Records
1*	25	129	12	3	62	23	1	73
2	35	96	13	14	111	24	5	71
3	42	80	14	13	112	25	4	63
4**	12/8	70/226	15	2	67	26	3	63
5	7	568	16	7	115	27	1	66
6	10	113	17	9	115	28	2	58
7	3	96	18	7	112	29	1	65
8	6	105	19	1	74	30	2	54
9	2	69	20	2	67	31	1	67
10	5	124	21	3	70	32	4	73
11	9	77	22	2	72			

* DJ01 contains 14 non-contemporaneous records written to memorialize older dreams, four of which were veridical, but did not have contemporaneous documentation.

** DJ04 values are split between dates prior to September 16, 1990, and those on or after that date. This is to allow analysis of the first 365 days recorded.

TABLE 2
Veridical psi Events from First Year of Dream Journal Records

Journal	Number of Records	OBE	Precognitive	Other	Veridical psi
1	129	3	19	3	25
2	96	11	20	4	35
3	80	31	9	2	42
4 (up to 9/15/90)	70	5	6	1	12
Total	375	50	54	10	114

Comparing this group of journals against other research into OBEs, it represents one of the largest single-source collections of veridical OBEs described in the literature. In the Holden (2009) literature review of veridical perception cases connected to NDEs, a total of 107 cases were found, the earliest of which was from the year 1882. In the dream journals that form the basis for this article, 114 veridical records occur in the first year, but this number is quickly surpassed and more than doubled as of the present day. No other study of OBEs reviewed for this article, whether connected to NDEs or not, involves a comparable number of veridical cases. On the contrary, a frequent complaint is the lack of examples in the literature (Ring & Cooper 1997).

Fifty-two percent of all dreams recorded in the first year are coded as "Uncategorized." This means that the dreams do not match the criteria of dream types of interest. Table 3 provides a breakdown of the categories of interest, and their incidence. The number of dreams in each group adds up to more records than contained in each journal because some records contain multiple scenes belonging to two or more categories.

When averaged over the first 365 days, precognitive and OBE dream types are nearly equal in incidence, and together they represent 34% of all dreams for the period. See Table 6, Item 11 of the Appendix, for frequency of dream types.

The high incidence of veridical records in DJ03 (52.5%) led me to conclude at the time that veridicality was a product of accurate records and the willingness to check them against potential real-world correlates. Another way to say this is that I stopped looking at veridical dreams as "special" and became fascinated by dreams that could not be verified. Many of those, ironically, were subjects that were impossible to verify, such as dreams of angels, heaven, OBEs with strangers, and spirit guides. For DJ03, once those dreams are also subtracted from the total, only 33.8% remain unaccounted for.

The number of veridical OBEs is closely linked to efforts made to verify them. During the making of DJ01, I had not yet learned to identify OBEs very well, and did not check for them as assiduously as I would later. While working on DJ02, I was aware that I had OBEs and had learned to check them, but had not fully committed myself to checking all of the dreams that appeared to be OBEs. Part of this had to do with the cost of long-distance telephone calls or the effort of writing long, unusual letters to friends and family who might be surprised by such communication. By the time I reached DJ03, I was sufficiently confident to attempt verification of more items than I had previously. The result of this, as seen in Table 3, is a steadily increasing number of veridical OBEs, followed by a precipitous

TABLE 3
Classification of Dream Types from 9/16/89 through 9/15/90

Dream Type	DJ01	DJ02	DJ03	DJ04	Totals
Uncategorized	80	48	27	40	195
OBE	1	0	0	0	1
Lucid OBE	4	4	2	3	13
Veridical OBE	2	7	30	5	44
Lucid veridical OBE	1	4	1	0	6
Precognitive	19	20	9	5	53
Lucid precognitive	0	0	0	1	1
Prophetic	3	0	0	1	4
Lucid prophetic	1	0	1	0	2
ADC	5	2	1	0	8
Clairvoyant	2	0	0	0	2
Reincarnation	2	4	1	6	13
Spiritual	3	3	5	6	17
Symbolic	0	0	0	1	1
Shared dream	0	3	1	0	4
UFO, aliens	6	0	0	0	6
Healing	0	0	0	1	1
Religious figures	0	1	2	1	4
Totals	129	96	80	70	375

drop starting with DJ04, when I stopped making an effort to confirm OBEs.

Characteristics of Veridical Records

When the journal was started, my focus was on precognitive dreams. I did not at first suspect that I might have recorded OBEs in the journal. Six weeks after starting the journal, I was on the phone with a friend who lived in California named Lisa Moore. Moore interrupted the conversation to ask if I'd dreamed about her recently. I thought I might have but wasn't sure and offered to check. She agreed, as if she expected I had dreamed of her, so I went upstairs to find the dream I had in mind.

I came back to the phone with my journal and read an account from about two weeks earlier. I hadn't written her name in the record, but one of the people in the dream reminded me of Moore and that was the basis for selecting the dream. Moore said it was a fair description of recent events in her life connected with the death of her cat during veterinary surgery after it was run over by a car. Impressively, the unusual detail of decapitation was included in my notes. As I learned after I had read the dream to Lisa, this is the first of three cases I know of where the person I dreamed about during an OBE actually saw me at their location. Because of this, Lisa had expected me to call and tell her of the dream. After two weeks of waiting for me to do this, I did call her, but not for the purpose of describing a dream. Frustrated that I hadn't volunteered any information concerning a dream about her, she gave up on waiting and asked me instead.

Moore's phone call alerted me to the possibility that when I dreamed of friends, family, or even strangers, I might be seeing events from their lives. This possibility contrasted with my expectation that dreams would be random or related to future events in my own life. To test this new hypothesis, I embarked on an effort to verify every dream that matched the following criteria:

1. It featured a person who ignored my presence,
2. I could identify at least one person in the dream, and
3. The dream contained unusual details that could be used for verification.

The first criterion may seem unusual, but it was my way of knowing that I was not a literal participant in the activity. In a typical OBE, I try to interact by talking to people, but they ignore me. Not infrequently, I assume they can hear me, but are purposely ignoring me. This causes me to become increasingly agitated as I make successively more aggressive attempts to force the people in the dream to acknowledge my presence. I learned over time that my observations from dreams like this could usually be confirmed by the subject of the dream.

In a small number of OBEs, not only am I aware that I am out-of-body, but the person I am trying to communicate with will simultaneously ignore me on a physical level but engage me telepathically to say that he is busy and cannot talk. An example of this is provided by a veridical OBE where I saw a person I knew, Dr. David Ryback of Atlanta, Georgia, talking to a tenant in the building they both worked in. Dr. Ryback's acquaintance was telling him how two cars he owned had been severely damaged on two separate occasions in the same week in the same way, by having tree branches fall on and crush their roofs. While continuing his conversation with this man, Dr. Ryback had a brief telepathic conversation with me on a different topic, the gist of which was that he provided a quick answer to a question I had and then told me he was occupied and could not communicate any further (Paquette 2012). This dream satisfied my third criteria for a veridical OBE because the detail regarding the accidents involving Dr. Ryback's acquaintance were sufficiently unusual and removed from my own knowledge to be described as anomalous.

Dream Characters

A dream character is simply any intelligent entity within a dream, whether it directly interacts with the dreamer or not (Waggoner 2009). Based on this, the "Dr. Ryback" from the dream described earlier would be a character created by my subconscious mind. Without claiming to know whether some characters are created this way, I doubt that all of them are, and suspect many are not. The reason is that veridical information implies a nonlocal source. Thus, my dream of witnessing Dr. Ryback's conversation is literally a viewing of the actual event, and "Dr. Ryback" is the person himself, not a subconsciously created manifestation. Whether this is a correct interpretation can be debated, but I do think it is objectively at least as credible as the "dream character" interpretation, on the basis of the many veridical examples available for review.

Spirit vs. Physical Content

Encounters in dreams with characters who engage me directly, particularly when they approach me or do something to get my attention, are a signal that the person is a spirit. When this happens, the dreams infrequently touch on matters that can be verified because they normally do not concern themselves with physical locations or living persons. If they are verifiable, they are usually precognitive or prophetic. "Precognitive" as used here means that I am shown something from the future. "Prophetic" is when I am told something about the future, sometimes without seeing it.

In comparison to the 226 veridical dreams contained in the first 32 volumes of the dream journals, 316 nonveridical dreams contain spirit-related content. For this article, a "spirit" is defined as the conscious element of a person's identity that survives death. Spirits belong to a person's identity whether that person is alive or deceased. This allows a distinction to be made between witnessing a person engaged in physical activity during an OBE, and communication with the spirit of a living person during an OBE. Because spirit-centric dreams rarely contain information that can be verified, it is easy to assume that the "spirits" who appear in them are dream characters rather than actual spirits. However, enough spirit-centric dreams are veridical that I find such a conclusion unsatisfactory.

For example, in a dream from August 12, 2003, the spirit of a recently deceased young woman gave me an urgent and disturbing warning for a relative of hers named James. James was a clerk who worked at an art supply store I shopped at occasionally. I had spoken to James on a handful of occasions while purchasing art supplies, but did not know him well. Though hesitant to pass on the warning to James, I did do it. James confirmed that his sister-in-law, with whom he was close, died within the last two weeks when her car was rammed by a police car during a high-speed car chase. He stated that she had appeared to him earlier that week in a dream and given him the same warning she had given me in my dream (Paquette 2011). In most cases, dreams that involve spirits concern people I cannot contact. In a very small number, such as this one, there is a reference to someone I do know or can find, and the dream may be verified. This is why it is important that an OBE includes someone I know or can find to contact for verification.

TABLE 4
Breakdown of OBE Types

Type	OBE Incidence 9/16/89–9/15/90	
OBE	1	1.6%
Lucid OBE	13	20.3%
Veridical OBE	44	68.8%
Lucid veridical OBE	6	9.4%
Total	**64**	**100.0%**

Lucid Dreams and OBEs

Many records describe events both as a "dream" and an "OBE." All OBEs recorded in my journals occur during dream states but not all dreams include OBEs. Veridical and nonveridical OBEs are sometimes but not always lucid. Therefore, some lucid dreams are also veridical OBEs (or other types of veridical dreams, such as precognitive or spirit communication). Table 4 shows the proportionate incidence of various OBE dreams.

78.1% of all dreams identified as OBEs in the first year of the journal are veridical. Lucidity, while not uncommon, does not ensure veridicality. The true incidence of nonlucid, nonveridical OBEs is likely higher than reported here, because such a dream has no factors that would aid in its identification as an OBE.

According to the literature, lucid dreamers can and do consciously alter the dream environment (Waggoner 2009). In OBEs, this does not occur. Another difference is that OBEs often begin with awareness of the dreaming state, but in lucid dreams this occurs after a dream is already in progress. Waggoner's observations about lucid dreams and OBEs match my own (Waggoner 2009:28–29), but his definition makes it difficult to use the word *lucid* to describe one's level of awareness within a dream.

Based on the distinction between an OBE and a lucid dream given by Waggoner, lucidity alone, or clarity of mind and awareness of the dreaming state, is not enough for a dream to be classified as "lucid." I prefer to describe lucidity as a cognitive quality that may be present within any dream rather than designating it as an exclusive characteristic of "lucid dreams."

The OBE Experience

It is tempting to say that waking from an OBE is somehow different from non-OBE dreams, but this is not true of the dreams recorded in my journals. Sometimes a particularly vivid dream will impress me to such an extent that it is uppermost in my mind for some time after I wake, but this is no different from other dreams. OBEs are sometimes vivid, sometimes not. Sometimes they are lucid, sometimes they are not. Some are very detailed, some are not. OBEs, whether veridical or not, do possess elements not found in other dream types but they are unrelated to vividness, urgency, or waking sensations.

In comparison to descriptions of OBEs by Robert Monroe (Monroe 1977), my own OBEs differ only slightly. Mine start in dreams, as do most of Monroe's (Tart 1998), but he describes very detailed tactile sensations that accompany leaving his body, such as feeling the grain of flooring beneath his bed as his astral body drifts through it. Whether this happens in my own experiences, I have no memory or record of it. Monroe describes spirit helpers who sometimes assist him during an OBE. I have made the same observation within my own, but his descriptions imply a different style of awareness (Monroe 1977, Paquette 2011). In his chapter on angels and archetypes, Monroe describes helpers as disembodied hands more often than as fully formed figures (Monroe 1977:127–135). When my records describe such helpers, they are always fully formed beings. However, like Monroe, I am sometimes aware of their presence without seeing them, as if they are "behind" me. Another difference is that my records describe spirit helpers who explain their business to me as they provide assistance. In Monroe's examples, the spirit helpers are more enigmatic and do not directly explain themselves (Monroe 1977).

Some of the earliest OBEs recorded in the journals involve a spirit guide who first tells me I am asleep, and then offers to teach me how to leave my body. In the first such dream, I was so exhilarated by the sensation of leaving my body that the spirit guide remarked it was "too soon" and that he would come back later when I was more ready. Before coming back to my body, I felt like I was being drawn upward into a huge tunnel along with other spirits. In later dreams I received training designed to teach me how to leave my body as a spirit. In several of these, I would feel a vibration such as Monroe describes, and then would become lucidly aware that my body was asleep. I would see my room, but also one or two spirit figures near my body as it lay in bed. They would take hold of me and literally pull me out of my body. In other OBEs, a spirit guide will direct another spirit guide to escort me back to my body. On arrival, there is a third spirit guide waiting for us, and they help me re-enter my body, after which I wake. Many of the OBEs that involve spirit guides are not veridical, though some are. However, the examples just given provide an idea what the process is like.

OBEs that do not involve spirit guides have their own characteristics. The first is that I will at some point within the OBE begin to feel exceedingly tired. This leads to a gradual collapse into "unconsciousness" within the dream followed by waking in my bed. In one veridical example from April 22, 1990, I observed my mother in her apartment, which was about 2,800 miles from where I was sleeping. I saw that she was on a date with someone and that they were listening to Schubert while she cooked something in her kitchen. While watching this, I suddenly became very tired and leaned into a wall opposite my mother's position in the kitchen. I then sank to the floor along it, making a kind of scraping noise against the wall. My mother suddenly turned to look directly at me as if alarmed, and then I woke. I called my mother later in the day and verified various elements of the dream. To my surprise, she said that she had been surprised while cooking that night by a strange sound coming from the wall opposite her. She said it sounded like a paper bag being scraped against the wall as it fell to the floor, followed by a thud, but she saw no source for the noise.

Immediately prior to waking from many dreams, I will find myself attracted to a small aperture at the end of a winding path. It could be a crack in a wall, a small hole, a crevice between two objects, but whatever it is it will start small and then become progressively smaller and more abstract as I am drawn into it. Very quickly it will become a series of convolutions, as if I am traveling down a long, cramped, twisting tunnel. After a short period of this, I will wake. This happens so often that I rarely bother to write it down any longer, but mention it to give a sense of how there is a distinct sense of physical separation and traveling to integrate one body with the other.

Identification of an OBE

The way I distinguish a probable OBE from other dreams is that the subject of the dream ignores me, and the context is a logical physical environment that is unaffected by my presence. If I am lucidly aware that I am out-of-body, then I need no other information to identify it as an OBE, but if I am lucid, the other information will be there as well. If the dream contains enough information to identify and contact a witness to the events in the dream, it becomes a probable veridical OBE. If it is subsequently verified by contacting one or more persons who were present in the dream, then it is counted as a veridical OBE. If it is not verified, it can be for one of several reasons. The witness may be mistaken, the description may not be adequate to inspire recognition, the witness may be misidentified, or the dream may literally be incorrect. I have not performed an analysis of this question, but can say that there are examples of each explanation in the journals. Witnesses do sometimes deny that a dream record is connected to them, but this is not common. A later study will attempt to fix a definite figure to this.

Very few OBE records contain a full sequence that includes the process of leaving the body, witnessing events at a distant location, and then returning to my body. No veridical record describes traveling from my physical location to the place where something is witnessed. This is unlike Monroe (1977), who writes of viewing terrain pass beneath him as he travels from one physical location to another during an OBE. As Table 5 shows, veridical OBEs in my journals only rarely include details about leaving or re-entering the body.

Table 5 covers the period of September 16, 1989, to September 15, 1990, the first year of the dream journal. Within that time frame, only 2 of 50 dreams include detail about the sensation of exiting the sleeping body

TABLE 5
Veridical OBE Records That Contain Information
about Exit, Re-entry, and Lucidity

Journal	Number of Records	OBE	Exit	Re-entry	Lucid
1	129	3	0	0	1
2	96	11	1	3	3
3	80	31	1	1	1
4 (partial)	70	5	0	1	1
Total	375	50	2	5	6

prior to a veridical OBE. Five dreams included detail regarding the return to the physical body after a veridical OBE. Six of the veridical OBEs were lucid. There is no significant correlation between the number of OBEs and the number of dreams containing information specific to exiting or re-entering the physical body.

The central question of inquiry into OBEs is whether the mind actually leaves the body. If it leaves the body, then brain-based theories of consciousness become untenable, and the concept of survival of consciousness is at least partially validated. If the mind does not separate from the body, but refocuses its attention elsewhere in what amounts to telepathy, an argument can be made that this does not conflict with brain-based theories of consciousness.

There are reports, and this includes my own observations, that some kind of movement from one discrete location to another does occur during OBEs (Monroe 1977, Paquette 2011). Whether or not this self-reported sensation is accurate, research into NDE-related veridical OBEs is highly suggestive that consciousness does not require a functional physical body (Lommel, Wees, et al. 2001). Studies of reincarnation (Stevenson & Samaratne 1988, Mills 1990, Haraldsson & Abu-Izzedin 2002) and mediumship (Rock & Beischel 2008) are suggestive of survival of consciousness after bodily death. In combination, these studies support the idea that consciousness is not dependent on the existence of a physical body. At the least, they provide a rationale for location-specific OBE travel.

Logistics

It took several months to learn how to record these experiences properly. Such simple logistical factors as buying a small flashlight and being careful with my handwriting were major technological advances in this effort. Of greater importance is that over time I learned that certain predictable errors were made while recording the dreams. For example, I might remember the appearance of something seen in an OBE but would often be incapable of understanding what it was or what its function was. To counter this, I learned to write descriptions without rushing to a conclusion about an object's purpose. Identification errors happened when I would mistake one person for another. My Uncle Tom and my friend Richard would be mixed up in dreams, but if I paid attention to the location of the dream, I could figure out which one of the two it was. Thanks to the large number of veridical OBEs involving both, I quickly learned how to identify them correctly and to spot identification errors with other people. It did not take long before I learned that identification errors were linked to specific people and conditions. With that knowledge, the number of errors decreased dramatically.

On any morning that I awoke from a dream, I first wrote down the information. After this, I normally described the dream to my wife and then either phoned, faxed, or wrote to the person involved to obtain verification. Most of the time telephone calls were enough, but in one particularly difficult example, tracking down the address of a "Richard" whom I hadn't seen in years proved elusive, so I started with a different Richard whom I could reach more easily, in the hope that it was an identification error based on the name the two men shared. After that didn't pan out, I again went looking for the original subject and after a month was able to obtain verification of the OBE from him. Most cases, however, were confirmed on the first attempt within hours of the OBE or by the following day.

Although my research was started in an effort to determine the truth of these phenomena and to get good contemporaneous records of them to test my hypothesis that memory errors on my wife's part accounted for her belief that I had experienced precognitive dreams, I did become convinced soon after beginning this exercise that the phenomena I found myself recording daily were common and had a nonlocal explanation. This means that I eventually hit a point where I was no longer interested in looking at the material as evidence for psi and began exploring the meaning of its more spiritual aspects. Practically speaking, this means that the highest percentage of veridical items are found in the first three journals, after which I ceased actively pursuing verification and paid more attention to my career as an illustrator and animator. Veridical items continued to appear, and still do, but these are verified passively for the most part, for I rarely make the effort to try to verify them any longer, having already acquired sufficient material to be satisfied. It is for this reason that I intend to confine my examples to the first years' worth of records.

Geography and Witnesses

Geographical factors are less important in NDE cases in which the subject of the NDE claims extra-physical perception of nearby objects, such as the operating theater occupied by the patient, the waiting room down the hall, or the site of an auto accident. Skeptics have asked whether

> . . . the sense of hearing [could] become a prime collector of data for the intuitive mind and the mediator for many apparently telepathic and para-psychological phenomena? (Wettach 2000)

The skeptical argument that the subject of an OBE has some kind of unconscious ambient knowledge of his immediate environment does not explain the material I will present because of the distances involved.

At the time the journal began, I lived with my wife temporarily in an apartment owned by her family in Manhattan. Within a few weeks we found an apartment of our own in Weehawken, New Jersey, and moved there. With the exception of the first couple of weeks in Manhattan and one trip to Miami Beach the following year, all of the records in the first three journals were made in Weehawken. This point is important because only once did a veridical OBE involve someone in Weehawken.

One dramatic OBE involved a friend and colleague of my wife's, named Joseph Fazecas, who lived in Manhattan. At the time, we lived in Weehawken. I dreamed that I visited Fazecas at the hospital. During the OBE, I was sure he had died. After describing this to my wife, she became alarmed and called her office, where they both worked, to check on him. He wasn't at work because a little earlier he'd had a serious heart attack and had been taken to the hospital for coronary bypass surgery. He survived the crisis, but—as an aside—I wonder if he had an NDE and if that is why I thought he was dead. In any case, the rest of my experiences involved people who lived much farther away. The closest was Brooklyn, but the most common locations were California, Minnesota, Florida, and Japan. In most examples, locations no fewer than 1,000 miles from where I was sleeping were involved.

The people who served as witnesses were friends, family, and acquaintances. This does not mean, and should not be inferred to mean, that I had such close ties with these people that I could make accurate guesses about the specific items recounted here. The nature of the information conveyed in the dreams did not lend itself to an explanation of being predictable even to the people concerned. This point is important when comparing these examples against other published cases in which the percipient may not have been expected to know anything about, for instance, hospital procedures, but those same procedures would be predictable to staff because they were part of a well-rehearsed procedure.

Documentation and Criticism

The principal value of laboratory-conducted OBE experiments is that documentation created in the context of experimental research can be more reliable than data related to spontaneous experiences. This conceit makes little sense in the context of the kind of criticisms leveled by skeptics against any data that tend to support a paranormal explanation. One common criticism is that a weak, addled, or confused state of mind contributes to memory errors on the part of patient and witness that becomes part of after-the-fact documentation (Evans 2002).

This is why before-the-fact documentation becomes valuable. It effectively eliminates all criticism of the "after-the-fact memory error" variety. Because all of my cases contain this kind of documentation, the memory error explanation cannot be credibly used. If one were to insist that memory error created false positive results, one would first have to discredit either the reliability or the provenance of the documentation. If that discrediting were accomplished, then the memory error explanation would be invalid or irrelevant.

In parapsychology circles there is a long history of criticizing the provenance, quality, interpretation, and credibility of documents. In the famous case of Nobel-Prize winning medical researcher Alexis Carrel, his documentation, eyewitness testimony, and reputation were all ignored by skeptics of his time when he described the case of a person he treated at Lourdes, who appeared to have been miraculously cured of tuberculous peritonitis (Moseley 1980).

More recently, skeptic Richard Wiseman has suggested that persons who believe they have experienced nonlocal sensory phenomena have misunderstood the connection between two distantly related things, do not understand probability, are fantasy-prone, or have poor cognitive abilities (Wiseman & Watt 2006). Even he has admitted, however, that there is little evidence to support the notion that people who claim to have experienced various types of psi events can be predictably described this way (Wiseman & Watt 2006:326–327).

Wiseman has not hesitated to find other means of criticizing studies that support psi hypotheses. According to Chris Carter, Wiseman has a habit of ignoring data that are not convenient to him. In reference to a replication of the "dogs that know" series of experiments conducted by Rupert Sheldrake, Carter wrote the following of Wiseman:

> Here we have a case in which Wiseman replicated a successful psi experiment, and then attempted to explain away his successful replication by arbitrarily ignoring most of his own data. (Carter 2010)

If Nobel-Prize winning scientists are not credible as first-hand witnesses, and documents produced by such witnesses are so fallible that colleagues will ignore them, and then later researchers such as Wiseman will ignore their own results to support their preconceived conclusions, then clearly no scientist in any discipline is safe from criticism. With this in mind, it must be accepted that all documentation, all researchers, and all witnesses are vulnerable to some kind of criticism, even if that criticism is ill-founded.

James Randi, the well-known skeptic, had this to say about me in a private correspondence with my friend Richard Breedon: "When I was a kid, I successfully predicted the outcomes of hockey games by having some 30 different letters notarized, each different from the others, and merely produced the correct one after the game" (Randi 1999). By this he meant to insinuate that I could have intentionally hoaxed Breedon. By extension, Randi's suggestion implied that even if I had notarized every page of the journal, a very expensive proposition for my limited means at the time, a critic could simply make the false claim that alternate notarized pages existed to demonstrate that notarization or any other form of proof can be manipulated into meaninglessness by a determined individual. The very same criticism can be leveled at any document created by any person for any purpose. In other words, at this level, the criticism is worthless because it can be suggested of anything.

Validity

This article is based on documents rather than memories, most of which are more than 20 years old. The reliability of the documents is something to which I can attest, having written them myself one entry at a time, on almost every morning between the first entry in 1989 and the present. Every OBE example is verified by at least one witness if not two. My wife was witness to the original entries,[1] as well as to my verbal elaborations of them, prior to the majority of any phone calls made for the purpose of verification. Secondly, the persons I called are witness to the fact that I told them my dream prior to them verifying any part of it. In some rare cases I obtained a verification letter from a witness. More often I simply recorded the results in the margin of the entry to which it belonged, along with the date of the verification, how verification was obtained, and from whom. I also noted differences or disagreements. I have discovered that witnesses sometimes do forget these incidents over time. For this reason it is good that the original margin notes have been preserved.

Three Examples

For evidential reasons, I will present only examples that have independent contemporaneous written verification. Scans of these documents are provided in Appendix 1. By limiting the examples in this way, numerous perfectly valid examples are ignored, many of which are at least as interesting as those presented.

As stated earlier, there are 50 corroborated veridical OBEs from the first year of the journals. Three notable examples involve my friend Richard Breedon, a high energy physicist who was working in Japan at the time of the first two examples, then at the University of California at Davis for the last.

Breedon provided written confirmation of all three items, though it is clear by his language that he is skeptical of a paranormal interpretation. Regardless, he confirms many key aspects of the OBEs. Briefly, these are the details.

OBE 1. On January 11, 1990, I dreamed that I was out-of-body and knew it. I visited Breedon in Japan, where I saw him sitting at a table

> ... doing something to these little wafers, or tiles. They have letters written on them and are a little bigger than scrabble pieces. He tosses them into the box when he is done with them. (Paquette 1990a:150)

When I called him—the first time I had ever made such a call—he confirmed that he had been doing just that for the past several days. He'd been cutting up pre-amplifier cards, labeled them with lettering that looked remarkably like the size and font used for Scrabble game pieces, and then placed them into a box on the table. Here is an excerpt from a letter he sent to corroborate the event:

> As a Research Physicist supported by the National Laboratory for High Energy Physics in Tsukuba, Japan, and also by the University of California, Davis, I am paid to be skeptical ... this is the first time [that the main purpose of one of Andrew's phone calls] was to recount a dream. He simply told me that he had just had a dream with me in it, with at first no indication that he might suspect that I would recognize details in it of my life or work in Japan (which, by the way, he has not seen, nor at that time had he seen any photographs).
>
> He said he pictured me working at a large desk, perhaps metal and grey. ... He saw some sort of machine in front of me, and that I was doing something with the machine to square or rectangular blocks. He said the blocks had letters on them, sort of like Scrabble blocks. After this, I placed the blocks in what he described as a trash container strapped to the side of the table. (Note: since I am writing this from memory, I may have a tendency to remember most clearly the details that correlated most closely with what I had been doing. He may have told me other details which did not correlate and I have forgotten.)
>
> It so happens that I had spent several hours earlier on the day he called, and also on the previous day, sitting before a large, white table, slicing electronic pre-amplifier cards with a cutting machine ... after this I got a piece of Teflon tape and wrote letters on the card to identify it (A, B, C, ..., AA, BB, CC, ..., AAA, BBB, CCC, etc.) with a black magic marker. I then placed the cards in a cardboard box sitting on the table.
>
> I was astounded by the accuracy of his description of what he had seen in his dream in also describing what I had been doing earlier ...
>
> In and of itself, this correlation, while significant, would not support a hypothesis of pre-cognition over more mundane explanations. (Breedon 1990a) (Item 3 in the Appendix)

Richard's mistake of confusing "pre-cognition" for what was, more accurately, an OBE was likely due to the fact that he was also aware of some precognitive dreams of mine and may not have been familiar with the concept of an OBE at the time.

OBE 2. On April 4, 1990. I dreamed of Breedon in Japan. Again I was aware that I was out-of-body. I saw him with a young Japanese woman in an interior space. They went to a car and pulled out a large object, which I illustrated in the journal. It was about as wide as the car, a foot deep, mostly black and white in color, and reminded me of the Manhattan skyline somehow. They then carried the object into his office.

Breedon initially said that he couldn't think of anything that matched these details (Item 4 in the Appendix), but a little later I received a letter from him saying that one of his colleagues noticed that my floor plan for the office was correct, though I put it on the wrong floor (Items 5 and 6 in the Appendix). Then he remembered an incident when one of the Japanese secretaries went to his car with him to help carry an electronic keyboard to his office so he could show it to colleagues there (Breedon 1990b). The alternating black and white keys of a keyboard have always reminded me of stylized representations of the Manhattan skyline, but Breedon did not understand this comparison. Figure 1 illustrates the graphic similarity between these two subjects.

OBE 3. On May 1, 1995, I dreamed that I visited Breedon and his wife Pat as she gave birth to twins in California. At the time I lived about 3,000 miles away, in Maine. I immediately sent an email to congratulate him (Paquette 1995) (Item 1 in the Appendix). He responded with the following message (Item 2 in the Appendix):

> Right you are! Born just hours before the time your message arrived here. How do you do it? I showed your message to two professors I work with. One said you had had to get it off the Internet (although I made absolutely no postings), the other simply said, "very good!" (Paquette 1995)

It should be noted that Pat gave birth six weeks prematurely and that I was not tracking her pregnancy, as Richard wrote in a communication to James Randi (Breedon 1999). Breedon read the message at his office on his way home from the hospital at about 5:00 a.m., prior to his having notified anyone that his children had been born. Because of the time stamp on the email, Breedon was able to ascertain that it was sent no more than three hours after his wife delivered the babies. On my end, I didn't send the email right away because the OBE was so realistic that I remembered it as if I had actually been there that morning when it happened. It was only a couple of hours after waking that I remembered that Davis, California, was 3,000

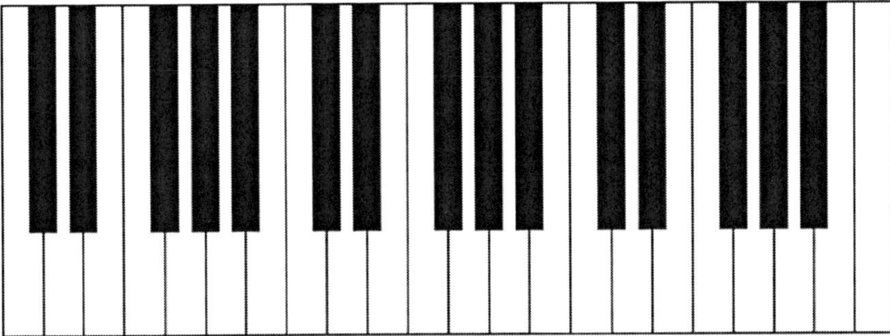

Keyboard layout

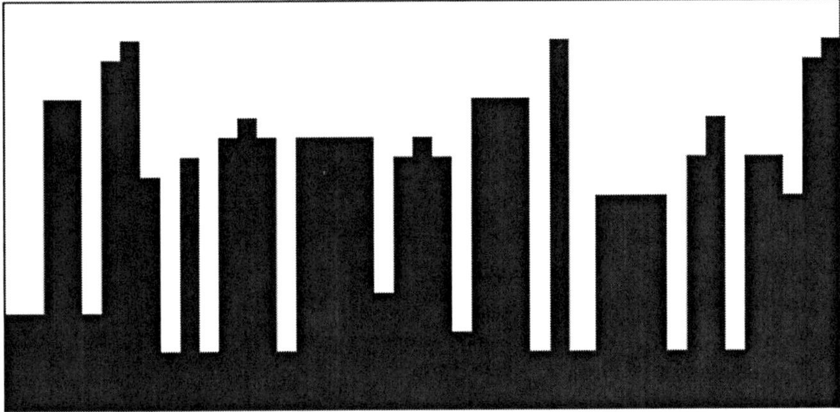

Stylized silhouette manhattan skyline

Figure 1. Comparison of keyboard layout and silhouette of Manhattan skyline shows clear relationship of alternating rows of long, thin, white and black rectangles.

miles away, so it had to have been an OBE. It was then that I rushed to my computer to congratulate him.

As stated before, "long distance" does not simply refer to information that is just outside of normal perception, such as a conversation that takes place a few rooms away from a recovery room in a hospital. These incidents are separated by geographically great distances.

OBE 4, a Detailed Example

On June 10, 1990, I had one of the clearest veridical OBE experiences found in my journals. It involved my uncle, Tim McGlynn, a man I barely knew at the time. The record contains five scenes, only one of which refers to McGlynn. This is the only item I checked. Here is that scene, extracted verbatim from the record:

> Aunt Terry and Uncle Tim in a room with a couple of paintings in it. One of them is talked much about. Tim takes pains to point out to me the Art Nouveau style trees at the horizon, stylized and flat. I thought at first he was saying that Terry had helped paint parts of it, but I think they were just talking about it and Tim made it by himself. (Pacquette 1990b)

Figure 2 shows the drawing from my dream book:

Figure 2. Scan of drawing from June 10, 1990, dreambook entry.

After waking, I decided to clean up this rough sketch from the journal and send it to my Aunt Terry in Minnesota by fax to see if it made any sense to her. It didn't make sense to me because to my knowledge McGlynn had never made a painting in his life. My Uncle Thomas Paquette was a painter, but not McGlynn. After making the cleaned-up drawing, I faxed it to my aunt's office (Item 7 in the Appendix) with a cover letter, then called to tell her of the dream. She surprised me by saying that McGlynn had recently been inspired by my Uncle Thomas and had finished his first painting just the day before. That very morning they discussed it because he wanted to hang it in the kitchen, but she didn't like the painting and didn't want him to. With that, I asked her to check her fax machine for the drawing I'd sent before calling. She came back to the phone, agreeing it was a remarkable likeness and that my cartoons had captured her reaction to McGlynn's

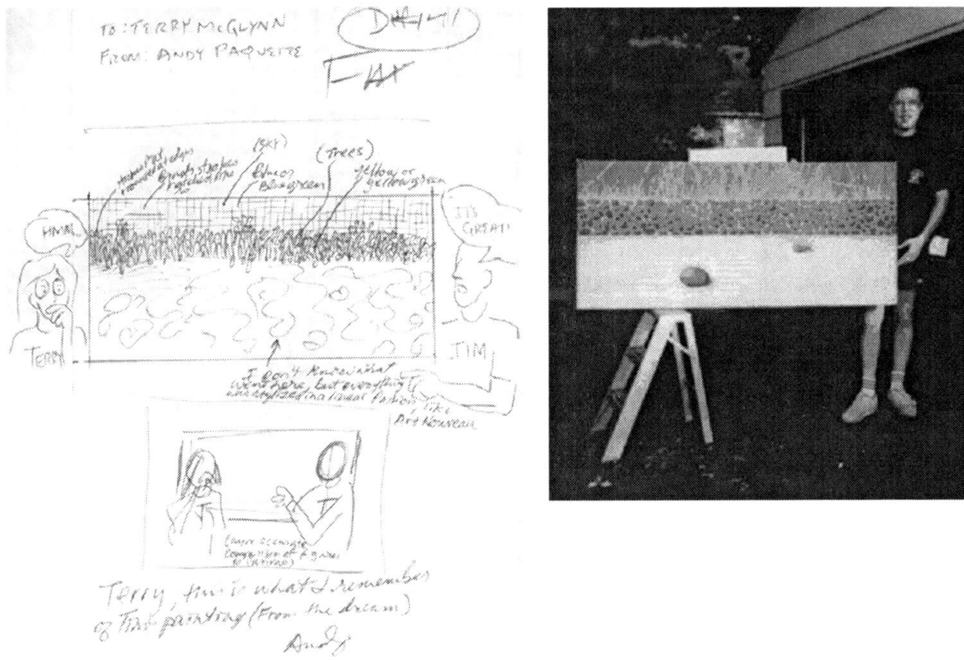

Figure 3. Revised, faxed version of drawing, and photo of actual painting (McGlynn 1990, Pacquette 1990b).

request and his enthusiasm that morning. Figure 3 shows the fax I sent (Item 7 in the Appendix), along with a Polaroid photo of the painting itself, sent by mail to me a little later (Items 8, 9, and 10 in the Appendix).

Discussion

The record of this OBE may not have many words, but the image is what makes this an impressive match to me, because the variety of possibilities available in a painting is enormous—far greater than for names of people or places. First, there are the dimensions, which are estimated at about 25″ × 50″. The proportions are exactly the same as the actual painting (a 1:2 ratio) and dimensions are close to or the same as what is shown in the photo. This aspect ratio is unusual for store-bought canvasses, which are most often either 2:3 or 3:4.

Second, there are the colors. All of the colors I mention in my drawing are not only present in the actual painting, but there is only one other color in the image: the browns of the rock wall. Third, it is a stylized image, which is another factor that makes the image unusual. In addition to these things, the composition is correct, and it is the right subject. It could have been anything from a dragon to a self-portrait, splatters of random color or cowboys. Instead, it is what I dreamed it to be, a specific kind of landscape.

There are two discrepancies, both of which are minor. The rock wall is transposed with the band of stylized bamboo, and the large gray rock in the foreground is absent. I say these are minor because, as a professional artist myself, I have made much greater errors when trying to remember a painting seen while wide awake just the day before. On that occasion, my wife and daughter had gone to Paris with me and visited the Louvre. While there, we saw many famous paintings, including Leonardo daVinci's *Mona Lisa*. The following day, after we'd returned home, all three of us attempted to draw it from memory. None of our drawings were any closer than the sketch of McGlynn's painting reproduced here is to the actual painting, though all were recognizably based on the *Mona Lisa*.

The examples involving Breedon involve very great distances, from New Jersey to Japan in the first two, and from Maine to California in the last. Breedon is an excellent witness and trained observer. He is also skeptical and went to the trouble of sharing accounts of these and other similar experiences related to me with James Randi (Breedon 1999). As Breedon told me later, he remains without an adequate or convincing explanation for the events described here, though he has thoughtfully considered many possibilities and has invited others to do the same.

Conclusion

Despite the apparent lack of very many examples of veridical OBE events in the parapsychological literature, the cases themselves do exist. It takes only one of these to cast doubt on the argument that human consciousness is brain-based. This may not appear to be adequately justified in a world that contains many mechanical devices capable of invisibly transmitting messages, such as cellphones, radio transmitters, and TV stations. However, veridical OBEs, whether connected to an NDE or not, sometimes also contain veridical information regarding communication from deceased persons, past-life memories, or information about the future (Stevenson & Samaratne 1988, Targ, Katra, et al. 1995, Rock & Beischel 2008, Paquette 2012). There is no mechanical counterpart to these last three types of information transmission because they involve information that, according to a nonparanormal explanation, is impossible to know.

The argument that consciousness is not brain-based has ramifications for researchers who attempt to provide nonparanormal explanations for such things as NDEs (Ehrsson 2007). It is my hope that the few examples provided here might provide some balance to this debate.

Note

[1] To satisfy a question from one reviewer, I will note that I did sometimes ask my wife to sign the journal record as accurate in her view, though I normally did this only for items that pertained to her rather than every entry she witnessed prior to verification, which was almost all of them.

Acknowledgments

Thanks to Janice Miner Holden and Rudolf Smit for their helpful suggestions while preparing this article.

References

Bonenfant, R. J. (2000). A near-death experience followed by the visitation of an "angel-like" being. *Journal of Near-Death Studies, 19*(2), 103–113.

Breedon, R. (1990a). Letter to David Ryback. D. Ryback and A. Paquette correspondence. Ibaraki-Ken, Japan: 3.

Breedon, R. (1990b). Keyboard dream 4/4/90. A. Paquette and K. Paquette correspondence. Ibaraki-Ken, Japan: 2.

Breedon, R. (1999). My friend Andy. Correspondence to J. Randi. Davis, CA: 8.

Carter, C. (2010). Heads I lose, tails you win, or how Richard Wiseman nullifies positive results, and what to do about it: A response to Wiseman's (2010) critique of parapsychology. *Journal of the Society for Psychical Research, 74,* 156–167.

Dell'Olio, A. J. (2010). Do near-death experiences provide a rational basis for belief in life after death? *Sophia, 49,* 113–128.

Dunne, J. W. (1927). *An Experiment in Time.* New York: Macmillan.

Ehrsson, H. (2007). The experimental induction of out-of-body experiences. *Science, 317,* 1.

Evans, J. M. (2002). Near-death experiences. *Lancet, 359,* 2116.

French, C. C. (2001). Dying to know the truth: Visions of a dying brain, or false memories? *Lancet, 358*(December 15), 2010–2011.

Greyson, B. (2000). Dissociation in people who have near-death experiences: Out of their bodies or out of their minds? *Lancet, 355,* 460–463.

Haraldsson, E., & Abu-Izzedin, M. (2002). Development of certainty about the correct deceased person in a case of reincarnation type in Lebanon: The case of Nazih Al-Danaf. *Journal of Scientific Exploration, 16*(3), 363–380.

Holden, J. M. (2009). Veridical Perception in Near-Death Experiences. In J. M. Holden, B. Greyson, & D. James, *The Handbook of Near-Death Experiences: Thirty Years of Investigation,* Santa Barbara, CA: Praeger/ABC-CLIO, 185–211.

Irwin, H. (1987). Out-of-body experiences in the blind. *Journal of Near-Death Studies, 6*(1), 53–60.

Lommel, P. v., Wees, R. v., et al. (2001). Near-death experience in survivors of cardiac arrest: A prospective study in The Netherlands. *Lancet, 358*(December 15), 2039–2045.

Long, D. J. (2010). *Evidence of the Afterlife.* HarperOne.

Lundahl, C. R. (2000). Near-death studies and modern physics. *Journal of Near-Death Studies, 6*(1), 53–60.

McGlynn, T. (1990). Confirmation photo and letter. A. Paquette correspondence. Minneapolis: 1+photo.

Mills, A. (1990). Moslem cases of reincarnation type in Northern India: A test of the hypothesis of imposed identification Part I: Analysis of 26 cases. *Journal of Scientific Exploration, 4*(2), 171–188.

Monroe, R. A. (1977). *Journeys Out of the Body*. Garden City, NY: Anchor Press/Doubleday.
Moody, R. A. (2001). *Life after Life*. London: Rider.
Moseley, J. (1980). Alexis Carrel, the man unknown. *Journal of the American Medical Association,* *244*(10), 1119–1121.
Paquette, A. (1990a). Dream Journal 1. *Dream Journals of Andrew Paquette*. Weehawken, NJ: 186.
Paquette, A. (1990b). D#141. T. McGlynn. Weehawken, NJ: 1.
Paquette, A. (1995). Congratulations. R. Breedon Correspondence. Portland, ME: 1.
Paquette, A. (2011). *Dreamer, 20 Years of Psychic Dreams and How They Changed My Life*. Winchester, UK: O Books.
Paquette, A. (2012). A new approach to veridicality in dream psi studies. *Journal of Scientific Exploration, 26*(3), 589–610.
Randi, J. (1999). Re-sending inquiry. D. R. Breedon Correspondence. Davis: 2.
Ring, K., & Cooper, S. (1997). Near-death and out-of-body experiences in the blind: A study of apparent eyeless vision. *Journal of Near-Death Studies, 16*(2), 101–147.
Rock, A. J., & Beischel, J. (2008). Quantitative analysis of research mediums' conscious experiences during a discarnate reading versus a control task: A pilot study. *Australian Journal of Parapsychology, 8*(2), 157–179.
Sabom, M. B. (1998). *Light and Death: One Doctor's Fascinating Account of Near-Death Experiences*. Grand Rapids, MI: Zondervan.
Schwaninger, J., Eisenberg, P. R., Schechtman, K. B., & Weiss, A. N. (2002). A prospective analysis of near-death experiences in cardiac arrest patients. *Journal of Near-Death Studies, 20*(4), 215–232.
Stevenson, I., & Samaratne, G. (1988). Three new cases of the reincarnation type in Sri Lanka with written records made before verifications. *Journal of Scientific Exploration, 2*(2), 217–238.
Targ, R., Katra, J., Brown, D., & Wiegand, W. (1995). Viewing the future: A pilot study with an error-detecting protocol. *Journal of Scientific Exploration, 9*(3), 367–380.
Tart, C. T. (1998). Six studies of out-of-body experiences. *Journal of Near-Death Studies, 17*(2), 73–99.
Waggoner, R. (2009). *Lucid Dreaming*. Needham, MA: Moment Point Press.
Wettach, G. E. (2000). The near-death experience as a product of isolated subcortical brain function. *Journal of Near-Death Studies, 19*(2).
Wiseman, R., & Watt, C. (2006). Belief in psychic ability and the misattribution hypothesis: A qualitative review. *British Journal of Psychology, 97*, 323–338.

APPENDIX

Item 1: Email sent from the author to Richard Breedon regarding birth of his twin sons

Note that all personal address information is either defunct or deleted.

From: MX%"74663.407@compuserve.com" 1-MAY-1995 05:24:18.35
To: BREEDON
CC:
Subj: Congratulations!

Return-Path: <74663.407@compuserve.com>
Received: from arl-img-2.compuserve.com by BREEDON (MX V3.3 VAX)
     with SMTP; Mon, 01 May 1995 05:24:15 PST
Received: by arl-img-2.compuserve.com (8.6.10/5.941228sam) id IAA11480; Mon, 1
     May 1995 08:23:38 -0400
Date: 01 May 95 08:21:14 EDT
From: Andrew Paquette <74663.407@compuserve.com>
To: "INTERNET:breedon"
Subject: Congratulations!
Message-ID: <950501122114_74663.407_BHW52-1@CompuServe.COM>

Last night I dreamed you had your twins, so I'm guessing you are officially a "dad" now.
Andy

Item 2: An email written by Dr. Richard Breedon to the author

FROM: INTERNET:
TO: Andrew Paquette, 74663,407
DATE: 5/1/95 6:38 PM

Re: RE: Congratulations!

Sender: breedon
Received: from   by dub-img-2.compuserve.com (8.6.10/5.941228sam)
          id SAA07755; Mon, 1 May 1995 18:33:51 -0400
From: <breedon
Received: by   (MX V3.3 VAX) id 20004; Mon, 01 May 1995
          15:34:09 PST
Date: Mon, 01 May 1995 15:33:56 PST
To: 74663.407@compuserve.com
CC: breedon
Message-ID: ·
Subject: RE: Congratulations!

Right you are! Born just hours before the time your message arrived here.
How do you do it? I showed your message to two professors I work with.
One said you had had to get it off the Internet (although I made absolutely
no postings), the other simply said, "very good!"

Mother and babies are doing fine. One came out vaginally, the other by
cesearean. They are in the neonatal intensive care, but only because
they are a little small. Tell Stev.
                                                      Thanks, Richard

Item 3:   Confirmation letter written to Dr. David Ryback and forwarded to the author by Dr. Richard Breedon

AMY Experiment, TRISTAN
KEK · The National Laboratory
for High Energy Physics
1-1 Oho, Tsukuba-shi
Ibaraki-ken, 305 Japan
BITnet: BREEDON@JPNKEKVX
HEPnet: KEKVAX::BREEDON
TEL 81 298-64-3513; 64-1171 ext. 6411
81-298-64-2196 (home)
FAX 81 298-64-3284
TELEX 3652-534
CABLE KEK OHO

June 1, 1990

Dr. Ryback
c/o Mr. Andrew Paquette
USA

Dear Dr. Ryback:

Since the time I met Andrew Paquette more than three years ago, he has told me of varied experiences of pre-cognition involving images appearing in his dreams. His tales are usually quite fascinating, and often remarkable. It is, of course, quite difficult to prove a pre-cognitive experience. I am directly aware of this from having worked with the late Dr. Wilbur Franklin at Kent State University in the 70's on experiments testing for pre-cognition and tele-kinesis with Uri Geller. (Let me point out that Mr. Geller failed to produce positive results on any of our tests when under controlled laboratory conditions.)

As a Research Physicist supported by the National Laboratory for High Energy Physics in Tsukuba, Japan, and also by the University of California, Davis, I am paid to be skepti-cal. But I do not believe in what the philosopher of science, Huston Smith, has termed "Scientism," that is, the refusal to consider as an explanation for observed phenomena any hypothesis to which the so-called Scientific method cannot be directly applied.

Dream images, being among the most subjective of all human experiences since they cannot be seen by anyone but the dreamer, are especially fraught with difficulty when one tries to relate them to waking images of one's own or of others. There is so much repetition and commonality in the day-to-day experiences of people living even miles apart that it would take a remarkable coincidence of details to demonstrate a correlation that would force one to consider a hypothesis as extreme as pre-cognition after eliminating all other possible explanations.

Andrew has asked me to recount a description of a dream he told me by phone to my apartment in Japan some months ago, and to tell how it correlated with work I had been doing over the previous few days. In and of itself, this correlation, while significant, would not support a hypothesis of pre-cognition over more mundane explanations. However, as I am aware that you are collecting reports of many such experiences of his, I will submit for inclusion.

Andrew had telephoned me on occasion before, but this was the first time wherein the main point was to recount a dream. He simply told me that he had just had a dream with me in it, with at first no indication that he might suspect that I would recognise details in it of my life or work in Japan (which, by the way, he has not seen nor at that time had he seen any photographs).

He said he pictured me working at a large desk, perhaps metal and grey (If any details of this account differ from his own, please consider his as the more authoritative as I am writing from memory and he took notes). He saw some sort of machine in front of me, and that I was doing something with the machine to square or rectangular blocks. He said the blocks had letters on them, sort of like Scrabble blocks. After this, I placed the blocks in what he described as a trash container strapped to the side of the table. (Note: since I am writing this from memory, I may have a tendency to remember most clearly the details that correlated most closely with what I had been doing. He may have told me other details which did not correlate and I have forgotten.)

It so happens that I had spent several hours earlier on the day he called, and also on the previous day, sitting before a large, white table, slicing electronic pre-amplifier cards with a cutting machine. The machine was a Sunhayato PC-300 HandCutter, which resembles an industrial-strength paper cutter in both appearance and function. The pre-amplifier cards were about 5 cm x 13 cm, and I was removing about 4 mm from the edge of both sides of each card (72 in total) with the cutter. After this I put a piece of teflon tape along the sliced edge and wrote letters on the card to identify it (A, B, C, ..., AA, BB, CC, ..., AAA, BBB, etc.) with a black magic marker. I then placed the card in a cardboard box sitting on the table.

I was astounded by the accuracy of his description of what he had seen in his dream in also describing what I had been doing earlier. The table he described seemed to be larger, and the container was not strapped to the table but sitting on top of it. Although in his description, the size of the cards was not clear to me, I was especially impressed that he saw letters written on them.

I have enclosed a photocopy of both sides of typical cards. The black, hand-written letters are the ones I was writing. I also enclose a Polaroid picture of the table in front of which I was sitting, with the cutting machine sitting upon it.

In the case you use any part of this report in a published account, I would appreciate being sent an advance copy of that portion to check for accuracy.

Sincerely,

Dr. Richard E. Breedon

Note that Breedon wrote this letter to Ryback at my request, because at the time Ryback and I were discussing the possibility of conducting a dream precognition experiment. We did carry out the experiment, starting on approximately March 31, 1990. Every morning I copied my dream journal record on my home copier and mailed the record to Ryback in a sealed envelope. The idea was to open them after something significant happened, and have the opening witnessed along with the postmark on the envelopes. Although many veridical dreams occurred during the experiment, most of them were of a personal, rather than newsworthy, nature, and Ryback felt this did not justify opening the envelopes. I later learned that Ryback had lost all but one of these envelopes in a move.

Item 4: The original fax sent to Richard Breedon regarding the second OBE

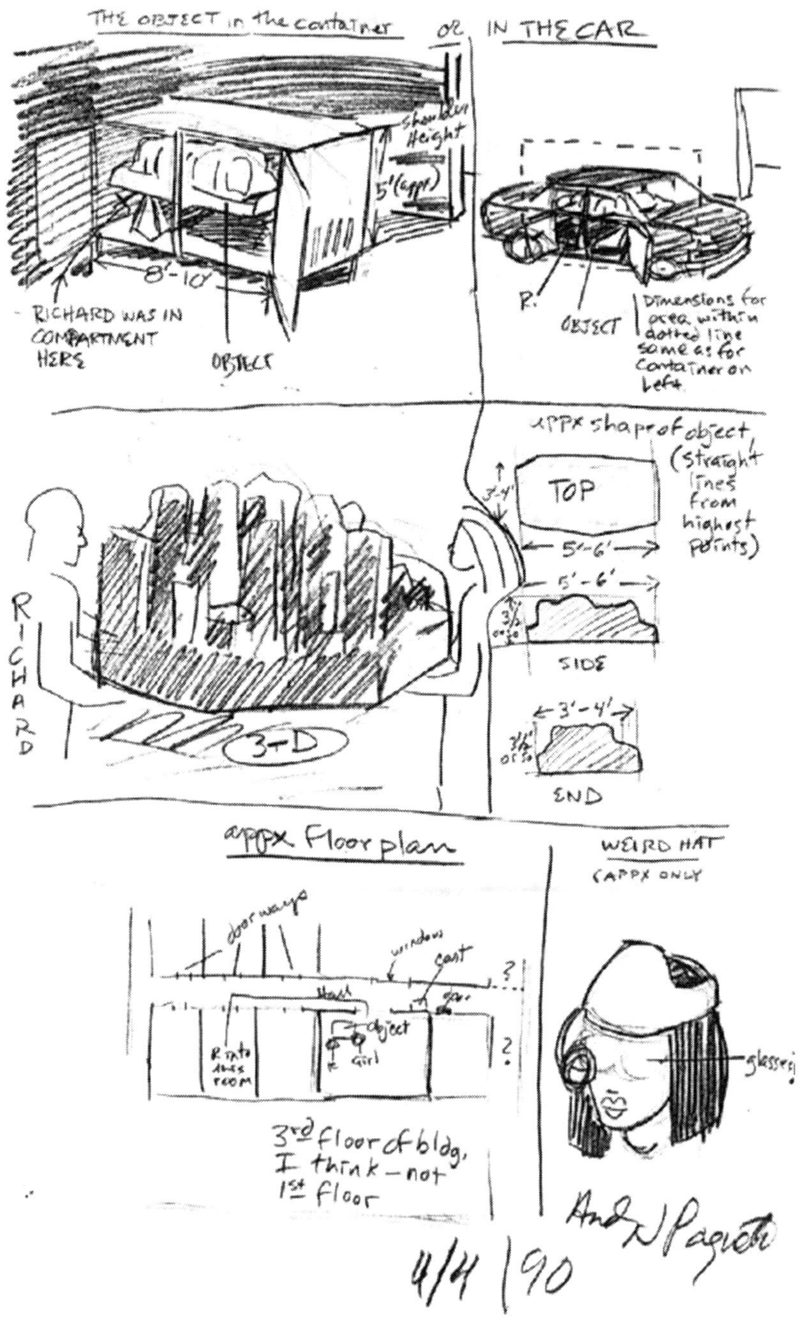

Item 5: Confirmation letter written to the author by Dr. Richard Breedon

Sorry I missed your visit here on 4 April. It would have been 3:30 in the afternoon in Japan. Your dream did not trigger any profound feelings of simultaneity, but a couple of details are worth noting. It was actually a student who works with me, Jeff, who noted that the floor plan that you drew has similarities to our offices. The room you put R. in, second from the end, corresponds to my own office, also second from the last on the 4th floor (while you said 3rd). On the opposite side of the hall you draw too many doorways (there is only one) but you put a window where there is indeed a window, along with several others, that look out into a center space of the building to allow light. Some months ago, one of the secretaries (Japanese, of course) helped me carry an electronic keyboard into the central office that I had just bought and had delivered, so that I could show her what it sounded like. It did not look like the skyline of Manhattan, but it was "black and white, mostly white" and came out of a large box. You drew the office where this occurred in approximately the right place but with many more offices separating mine from that one than shown. (If you can travel about in space, why not time?, the inquiring mind asks.) My secretary, however, was not wearing a funny hat, nor were there any unusual sightings made at the beginning of this month, but one must keep in mind that this whole place could be regarded as an unusual sighting. In conclusion, I would not want you to claim this as an unambiguous victory, but there are some interesting coincidences.

Item 6: The envelopes that contained each of the Breedon letters

Addresses are currently invalid.

Item 7: The original fax sent to Terry McGlynn

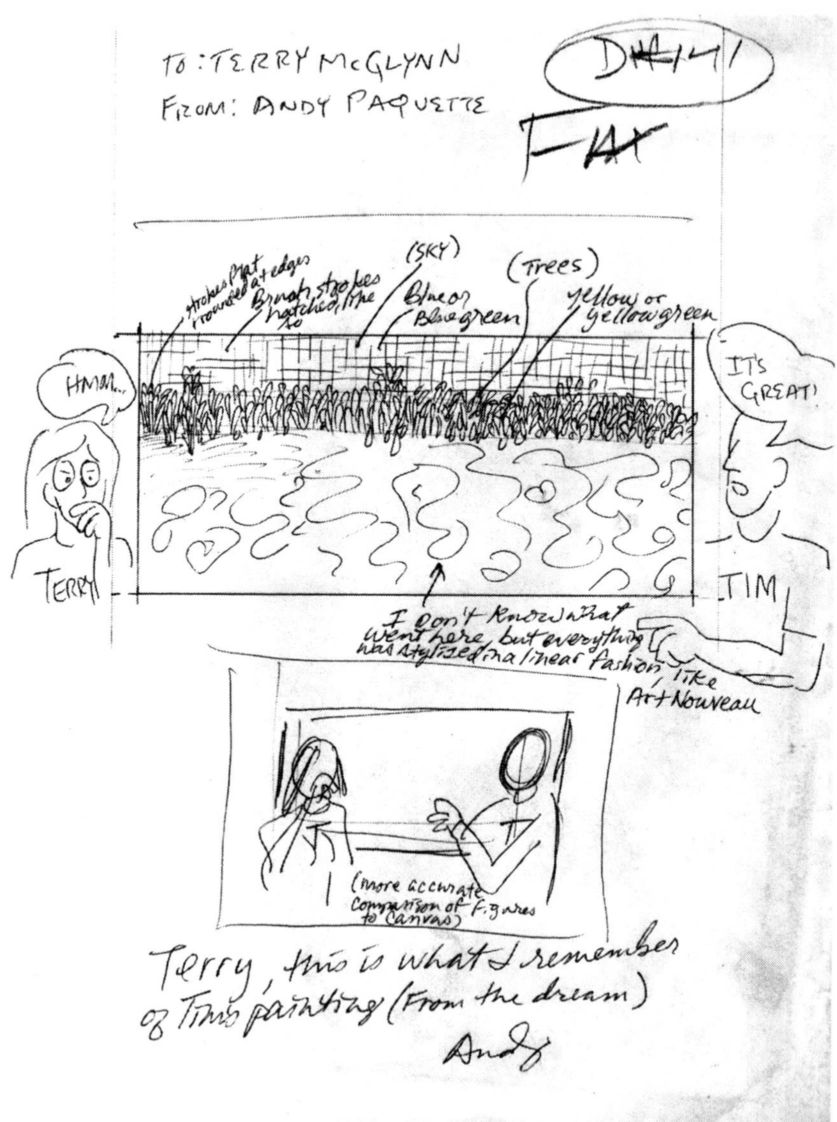

Items 8 & 9: Confirmation letter mailed to the author from Timothy and Terry McGlynn, and the envelope used (for dating)

The letter is addressed "Dear Uncle Andy . . . and uncle Kitty . . . " This is a joke on my uncle's part. Both addresses on the envelope are no longer valid.

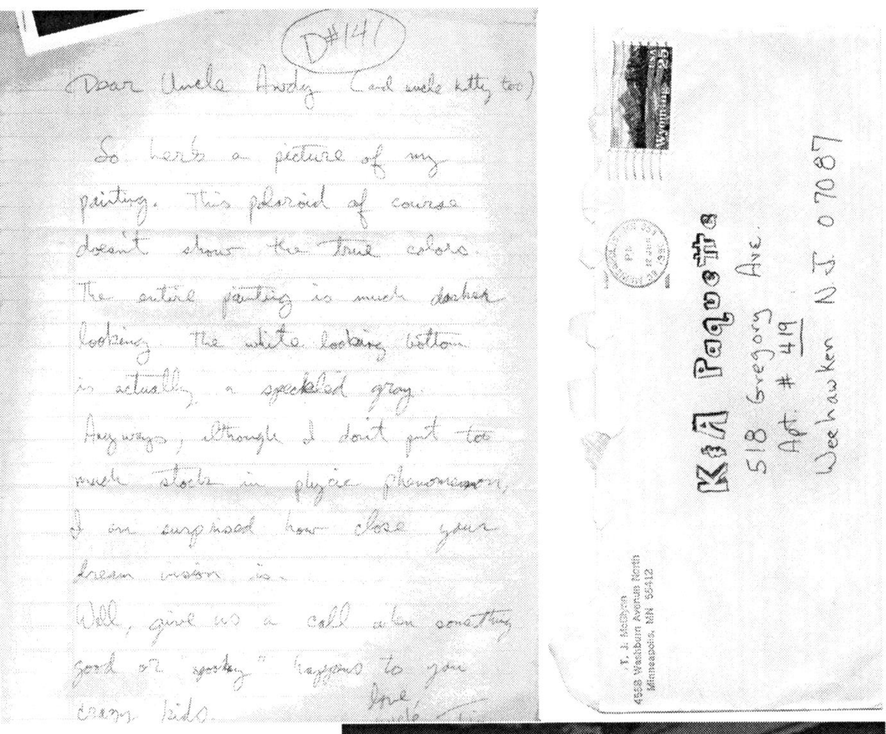

Item 10: The photograph attached to the McGlynn letter

Item 11: Frequency table for the 365-day period from 9/16/89 to 9/15/90

TABLE 6
Dream Types of Interest for 9/16/89–9/15/90

Dream Journal #			Frequency	Percent	Valid %	Cumulative %
1	Valid	ADC	3	2.3	2.3	2.3
		Clairvoyant	2	1.6	1.6	3.9
		Lucid OBE	4	3.1	3.1	7.0
		Lucid prophetic	1	.8	.8	7.8
		Lucid veridical OBE	1	.8	.8	8.5
		OBE	1	.8	.8	9.3
		Past era memory	2	1.6	1.6	10.9
		Precognitive	19	14.7	14.7	25.6
		Prophetic	3	2.3	2.3	27.9
		Spirit vision	2	1.6	1.6	29.5
		Spiritual mundus alo	2	1.6	1.6	31.0
		Spiritual mundus limus	1	.8	.8	31.8
		UFO, aliens	6	4.7	4.7	36.4
		Uncategorized	80	62.0	62.0	98.4
		Veridical OBE	2	1.6	1.6	100.0
		Total	**129**	**100.0**	**100.0**	
2	Valid	God/Jesus/Satan	1	1.0	1.0	1.0
		Lucid OBE	4	4.2	4.2	5.2
		Lucid veridical OBE	4	4.2	4.2	9.4
		Past era memory	4	4.2	4.2	13.5
		Precognitive	20	20.8	20.8	34.4
		Shared dream	3	3.1	3.1	37.5
		Spirit vision	2	2.1	2.1	39.6
		Spiritual mundus alo	2	2.1	2.1	41.7
		Spiritual mundus limus	1	1.0	1.0	42.7
		Uncategorized	48	50.0	50.0	92.7
		Veridical OBE	7	7.3	7.3	100.0
		Total	**96**	**100.0**	**100.0**	
3	Valid	God/Jesus/Satan	2	2.5	2.5	2.5
		Lucid OBE	2	2.5	2.5	5.0
		Lucid prophetic	1	1.3	1.3	6.3
		Lucid veridical OBE	1	1.3	1.3	7.5
		Past era memory	1	1.3	1.3	8.8
		Precognitive	9	11.3	11.3	20.0
		Shared dream	1	1.3	1.3	21.3
		Spirit vision	1	1.3	1.3	22.
		Spiritual mundus alo	4	5.0	5.0	27.5
		Spiritual mundus limus	1	1.3	1.3	28.7
		Uncategorized	27	33.8	33.8	62.5
		Veridical OBE	30	37.5	37.5	100.0
		Total	**80**	**100.0**	**100.0**	
4	Valid	God/Jesus/Satan	1	1.4	1.4	1.4
		Healing	1	1.4	1.4	2.9
		Lucid OBE	3	4.3	4.3	7.1
		Lucid precognitive	1	1.4	1.4	8.6
		Past era memory	5	7.1	7.1	15.7
		Precognitive	5	7.1	7.1	22.9
		Prophetic	1	1.4	1.4	24.3
		Reincarnation	1	1.4	1.4	25.7
		Spiritual mundus alo	5	7.1	7.1	32.9
		Spiritual mundus limus	1	1.4	1.4	34.3
		Symbolic	1	1.4	1.4	35.7
		Uncategorized	40	57.1	57.1	92.9
		Veridical OBE	5	7.1	7.1	100.0
		Total	**70**	**100.0**	**100.0**	

Journal of Scientific Exploration, Vol. 26, No. 4, pp. 825–853, 2012   0892-3310/12

Resonance between Birth Charts of Friends: The Development of a New Astrological Research Tool on the Basis of an Investigation into Astrological Synastry

GERHARD MAYER

MARTIN GARMS

Institut für Grenzgebiete der Psychologie und Psychohygiene, Freiburg im Breisgau, Germany
mayer@igpp.de

Submitted 10/19/2011, Accepted 7/31/2012

Abstract—In traditional astrological frameworks of interpretation, resonances between positions of astrological planets in the birth charts of friends (called "candidates" and their "partners") are assumed to play a decisive role. In the study presented here, this general claim is investigated at different levels of sophistication. For this purpose, five main hypotheses are formulated, all of which are different versions of the general assumption that there are more resonances between birth charts of friends than can be expected randomly. The material on which the study is based is taken from a questionnaire concerning the dates of birth of candidates to whom the questionnaire was distributed, as well as those of their partners. Having gained interesting results with partially supporting evidence, but also with elements that did not support the hypotheses, the experiment was repeated with a second sample. It failed to replicate the results of the first experiment.

Keywords: astrology—synastry—astrological resonance—friendship—astrological aspects

Introduction

Scientific efforts to investigate the correlation of astronomical factors and terrestrial events have not yielded much convincing evidence for the astrological hypothesis ("as above, so below"). Occasionally there are, indeed, some studies with "positive results" claiming supporting evidence. However, they often turn out to be methodologically flawed or based on incorrect assumptions, as is the case, for example, in the majority of studies investigating correlations between astrological signs of the zodiac and for example personality traits, occupational choice, and so forth.[1] Thus, the situation does not show great promise for research aiming to produce evidence for the astrological hypothesis (see Dean 1977, Kelly 1979, and Eysenck & Nias 1982 for comprehensive overviews of many older studies. A statement in accord with this by five prominent researchers in astrology and which takes more recent research into account is given in Phillipson (2000:142ff). Dean, Mather, and Kelly (1996:71–77) provide a meta-analysis of astrological studies).[2]

These discouraging facts are in opposition to the subjective (personal) evidence of many academically educated people who analyze horoscopes astrologically.[3] Thus there are, on the one hand, experiences of (subjective) evidence based on the application of experience-based assumptions (the application of traditionally handed-down rules of interpretation, of which some have been used successfully for hundreds of years),[4] and, on the other hand, almost every attempt to prove these rules has statistically failed to date. Skeptics perceive this as an indication of psychological processes (false attributions, the Barnum effect, etc.) playing a decisive role in the formation of experiences of astrological evidence (Dean 1999). This is a crucial point, of course, which has to be considered. However, it concerns mainly those kinds of astrological experiments that focus on the astrologer–client interaction, i.e. the interpretations of horoscopes as well as their communication (by the astrologer) and the understanding of them (by the client). Accordingly, the affirmation by a client that a horoscope interpretation is appropriate and consistent meets no reliable criterion for the validity of the astrological hypothesis (cf. Niehenke 1987:98–99, Dean & Mather 1994:16). As a consequence, choosing empirical facts (e.g., career choice, suicide, car accidents, (bad) luck in love, etc.) for testing this hypothesis seems to be a better approach. Such an approach has been chosen by many researchers, and much literature on this topic has been written (cf. Dean 1977, Eysenck & Nias 1982). Two kinds of research methods have been, generally speaking, applied to such investigations. The first is matching experiments, i.e. astrologers have to assign horoscopes to corresponding people (e.g., Steffert 1983, Böer, Niehenke, & Timm 1986, Dean 1986, Nanninga 1996/1997, Ertel 1998). This method is very time-consuming and strongly dependent on the skills of the astrologers concerned. The samples usually remain small. The second method often extracts single astrological factors from the overall context of the chart to test them for correspondences with empirical facts such as those given above (e.g., Gauquelin 1978, 1983, 1988a, 1988b, Niehenke 1987, Kollerstrom & O'Neill 1992, Sachs 1998, Denness 2000, Ruis 2007/2008). The approaches do not give consideration to the complexity of the astrological context of a chart and the symbolic ambiguity of the astrological factors (Koch 2002). Furthermore, some of the investigated "facts" such as personality traits or statements on (bad) luck in matters of love are based on self-characterizations (e.g., Smithers & Cohen 1982, Steffert 1983, Dean 1985a, 1985b, 1986, Niehenke 1987). As demonstrated by various studies, a rudimentary knowledge of the characteristics of one's own astrological sign has an influence on self-characterization, for example in questionnaires on personality (Eysenck 1981, Eysenck & Nias 1982:50–60). In addition to this possible bias, the characterization of one's own behavior, feelings, and perceptions which is often asked for in such studies confronts the researcher with possible discrepancies of self-perception, self-portrayal, and actual behavior of the participants which are based on insufficient self-reflexivity, projections, etc. (cf. Klein 1988 on this topic).[5]

Considering these problems, we developed a new research design that attempted to avoid some of the main weak points of previous studies:

- The complexity of astrological charts should be considered more adequately by not focusing solely on single constellations, and thereby taking the multiplicity of astrological aspects into account.
- A quantitative method should allow bigger samples than those possible with the usual matching test design.
- Problems occurring with the operationalization of psychological characteristics of participants should be largely avoided by avoiding verbal self-characterizations or characterizations by others. Instead, biographical facts should be elicited.

Synastry as an Object of Research

To take the last of the above-mentioned points into account (asking for biographical details), we directed our attention toward the field of friendships and long-term relationships. Personal relationships of this kind are highly personality-driven biographical facts, in contrast to more culturally determined biographical data (e.g., school-leaving exams); they are established on the basis of complex conscious and unconscious dispositions; we assume that the *engagement* in those relationships is to a lesser extent affected by self-perception than is the case with statements about one's own personality traits or about the subjective assessment of the quality of one's own relationships. As a rule, it also takes at least a little engagement of the related person of someone to call him or her a friend or lover. Therefore two people must be involved to establish a friendship or long-term relationship whereas self-characterizations (as gathered with personality inventories) depend strongly on self-perceptions which can be strongly biased (see above). Thus, one deals with biographical facts that can easily be elicited, and which need little interpretation: A friendship exists or, respectively, existed or not. However, the assessment of the *quality* of the relationships itself is not objective.

In astrology, the technique of synastry is fundamental to the analysis of personal relationships. Using this technique, astrologers make statements regarding the quality and intensity of the interpersonal and interpsychical dynamics between two people (e.g., Davison 1983, Meyer 1976, Arroyo 1978). One can even find this technique, in a rudimentary form, in the work of the ancient astrologer Ptolemy (Ptolemy & Ashmand 1822:124–128). Several scientific studies on synastry exist—one need only think of the famous astrological experiment made by Carl Jung (Jung & Main 1997:109–118). The results were inconsistent (Müller 1957, Eysenck 1983a, 1983b, Shanks 1983, Kuypers 1984, Shanks & Steffert 1984, Klein 1988, O'Neill 1986, 1989, 1990, 1995, Ruis 1993/1994, 1994, 1994/1995) but nevertheless had at least some promising approaches (Ruis 1993/1994, 1994, 1994/1995).

According to the doctrine of astrological synastry, the qualities of a relationship between two people are largely reflected in the interaspects of the two birth charts.[6] We may clarify this with the following example using astrological assumptions: If the zodiacal position of Saturn in the first chart is the same as the position of the Moon in the second one (astrologically signified as the constellation "Saturn in conjunction to Moon"), the first person will have a relatively high influence on the mood of the second person. The birth chart can be interpreted as a diagram of an individual "frequency response system" which is based on the different periods of revolution of the planets, and which can be seen, in some respects, as an analogy to a chord in music. We are calling the "superposition" of the "frequency response systems" of two people *astrological resonance*, in accordance with musical or physical phenomena. Thus two related people with a high resonance have many interaspects, whereas a low resonance is characterized by relatively few interaspects between the two charts. The accuracy of an interaspect contributes to the level of the resonance, and a higher accuracy causes a higher resonance. In our opinion, the aspect of class plays a minor part in this conception. Most of the previous investigations into astrological synastry focused on married couples (e.g., Jung & Main 1997, Van de Moortel 1998). We did not want to limit our study to this kind of relationship defined by a formal criterion (although it would be a "hard fact") because of our assumption that there is a significant number of married couples who have motivations for the relationship other than an attraction on a psychical level. This concerns marriage data from many years ago (e.g., such as that used by

Gauquelin) above all. If astrological interaspects actually correlate with relationship motivations, the level of the astrological resonance of friends and long-term partners should therefore be higher than the resonance between chance pairs of people. That was the basic assumption of our research project. In addition, we assumed that this effect will increase with those relationships that are characterized as *particularly intensive* by the participants.

Method

Relationships Questionnaire

We developed a questionnaire to collect the birth dates of the participants (referred to as *candidates*) and their friends and long-term partners, respectively (referred to as *partners*). A particular paragraph was devoted to the individual understanding of the quality of a relationship, that is the question of how a close friendship is characterized. We phrased our criteria broadly because people have very different attitudes concerning this matter, and excluded only relatives. In addition to the dates of birth of the partners, the candidates had to specify the sex of each partner, and to note if they characterized a relationship as particularly intensive. Some of our colleagues raised doubts as to whether certain candidates' previous astrological knowledge could lead to a particular partner choice strongly based on astrological criteria with the result that our data would be biased in a substantial way. Although the consideration of astrological criteria may be crucial for the partner choice by some strong adherents of astrology, we were of the opinion that this would not systematically bias our data. The instruction for the participants remained vague regarding the investigated hypotheses, and our method of data processing did not allow the outcome to be estimated. Therefore the probability that an intentional manipulation of the provided birth dates or selection of partners (e.g., in order to support the astrological hypothesis) would bias the outcome was certainly negligible.[7] Nevertheless, to take this point into account, we added two further questions in order to give an indication whether the candidates already had astrological knowledge, namely: *"Have you already read a book on astrology?,"* and *"Is the Sun sign an important factor for you regarding the choice of your friends?"*

In a second phase of data collection as part of the conceptual replication using the Internet, we modified the questionnaire slightly and added a second part for exploratory purposes. The modification of the first part concerned the indication of previous astrological knowledge and included three statements to be answered with "yes" or "no":

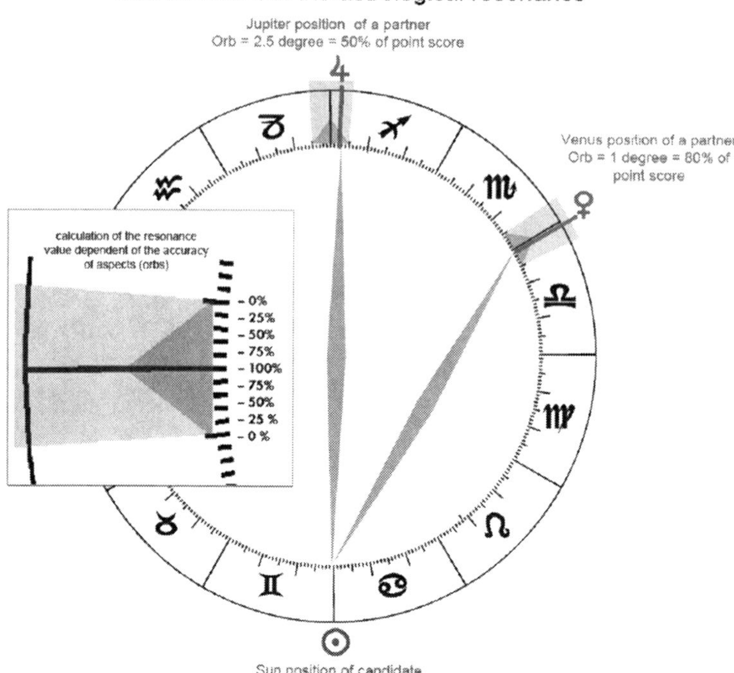

Figure 1. Quantification of the astrological resonance *R* between the candidate's Sun and two of the partner's planets.

a)   *I know the properties of my Sun sign well,*

b)   *I know my ascendant,* and

c)   *I know the zodiacal position of the Moon at the time of my birth.*

The second, optional, part of the questionnaire was accessible only when the first part had been completed. It provided the opportunity to give information on the kind of relationships (love, work life, friendship, recreational activities), their importance, their duration, and the frequency of contacts. Furthermore, information could be given about different aspects of life with regard to the relationships (erotic/sexual, emotional closeness/intimacy, good conversations, shared ideals, joint activities, learning from the partner).

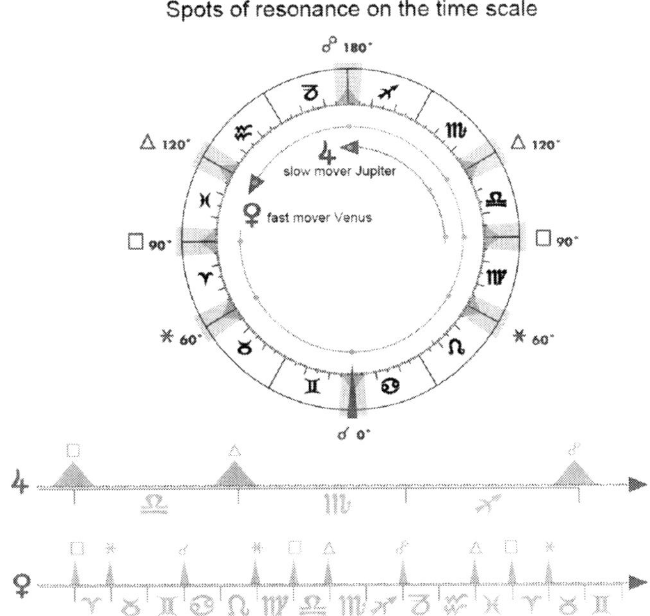

Figure 2. Points of resonance in the zodiacal circle that have been considered in our study (aspects).
Shown are two time scales of Jupiter and Venus which exemplarily demonstrate the different velocities of the planets Jupiter and Venus, and with that the different numbers of points of resonance during a period of time.

Assessment of the Astrological Resonance

We quantified the astrological resonance (R) by adding the weighted interaspects of the planets of the candidates (C) and partners (P). We chose the main astrological aspects of 0, 60, 90, 120, and 180 degrees, with an orb of 5 degrees. All planet pairs are considered equally. The weighting was linear: An exact aspect of two celestial bodies (planets plus Sun, in the following referred to as *planets*) received a value of 100 points. The value decreased linearly with an increasing orb and was given a value of 0 points at an orb of 5 degrees (see Figure 1 and Figure 2). The individual value used for the respective statistical operation depended on the selection of the planets, which varied according to the hypothesis.

Hypotheses

We formulated five hypotheses with sub-hypotheses that all are variations of our basic assumptions, and, therefore, were highly dependent. The first hypothesis concerned all of the "classical" planets of the candidate (except for Uranus, Neptune, and Pluto) and all of the "classical" planets of the partner (except for the Moon and the Trans-Saturn planets):[8]

1. ***All candidate planets (except Uranus, Neptune, Pluto), all partner planets (except the Moon) (H 1):***
   There is no more resonance than to be expected randomly between all planetary positions of the candidate and all planetary positions of the corresponding partner.

We limited the number of planets of the candidate to the Sun, Moon, Mercury, Venus, and Mars.

2. ***Individual candidate planets (prespecified), all partner planets (except the Moon) (H 2):***
   There is no more resonance than to be expected randomly between positions of the Sun, the Moon, Mercury, Venus, and Mars of the candidate and all planetary positions of the corresponding partner.

With the third hypothesis we directed our attention to the positions of particular planets of the partners. Saturn, for example, is said to be of particular importance relating to persistent and deep relationships:[9]

3. ***All candidate planets, individual partner planets (prespecified) (H 3):***
   There is no more resonance than to be expected randomly between all planetary positions of the candidate and the position of Saturn of the corresponding partner.

The fourth hypothesis refers to a widespread astrological concept, that most people are searching for a specific "planetary quality" in relationships: One may, for example, look for, above all, structure and trust in relationships—a quality of Saturn—whereas another may focus their attention on mainly aspects of harmony and hedonism—a quality of Venus—etc. Thus, we chose the individual planet of every candidate which had the largest resonance with his partner's planets:

4. ***Individual candidate planets (not prespecified), all partner planets (except the Moon) (H 4):***
   There is no more resonance than to be expected randomly between

the "most resonant" planet of the candidate and all planets of the corresponding partner.

The significant element in this hypothesis is that the "most resonant" planet of the *candidate* is not prespecified. The hypothesis refers to the significance of the resonance value of this "most resonant" planet, no matter which one it is. In order to determine the "most resonant" planet, we computed the averaged R of every candidate's planet to all partners' planets. The "most resonant" planet is the candidate's planet with the highest ranking (see Mathematical Appendix, Equation 2).

With the fifth hypothesis we refer again to the concept of a specific "planetary quality" in relationships, but we change the perspective, looking at the individually preferred planet of the *partners*. First, we computed the resonance of every partner planet to the "classic" planets of the candidate. Then we averaged the single resonance values of every partner planet. We chose the individual planet of the partners with the highest mean resonance value regarding all the "classic" planets of the candidate for statistical computation (see Mathematical Appendix, Equation 3):

5. ***All candidate planets (except Uranus, Neptune, Pluto, individual partner planets (not prespecified)) (H 5):***
   There is no more resonance than to be expected randomly between all planets of the candidate and the "most resonant" planet of the corresponding partner.

As mentioned above, we assumed that relationships which are characterized by the candidates as *particularly intensive* would correlate with higher resonance values than for other relationships. This assumption was the basis of the five sub-hypotheses:

Intense relationships do not show significantly higher resonance values than less intense relationships.

Assessment of Chance Expectancy (Reference Values)

The problem of estimating correct expectation values of complex astrological constellations adequately is not insignificant.[10] Using control groups is appropriate if distributions in the general population are known. This is not the case with complex astrological constellations.

O'Neill (1986, 1989, 1990) deals with this problem in his studies on synastry with eminent married couples by using another kind of "control." He built his "control" couples by "recomposing" birth data from his database taking the same age of the new partners (approximately, and taking the mean age difference of the whole sample of pairs into account). In a similar way, Ruis (1993/1994) built new pairs in his re-analysis of the Gauquelin data of married couples by combining all male subjects with all female subjects considering specific age-groups, likewise in order to estimate the theoretically expected aspect frequencies. In a second step, Ruis (1994, 1994/1995) used a "shifting method" suggested by Ertel to eliminate possible artifacts associated with the age-difference between partners.[11] With this technique, the birth dates were shifted by a constant number of days in both directions. The results were in the expected direction, that is the grand total of synastry aspects decreased with the increasing shift of the birth dates—but only for the first days of shifting. Ertel recommended this "shifting method" to us to avoid the problem of having to estimate the expectation values of complex constellations (personal communication). The candidate's date of birth and all of the partners' dates of birth should be stepwise shifted on the time scale (see Figure 10 in the Mathematical Appendix).

In our opinion, this method is not suitable for complex astrological constellations because it assumes a homogenous astrological time line. It follows the idea that the constellation actually found is a "perfect" constellation. Accordingly, the shifting of the dates of birth should lead to a lesser degree of R—comparable with the astrological idea of decreasing effect of an aspect corresponding with its increasing inaccuracy (orbs). This argument is flawed because it is based on the idea of a homogeneous time scale. This is not the case. We do not have an interval-scaled dimension which unproblematically allows diverse mathematical operations. That is to say, every point on the time scale has its own resonance value in relation to every other point, which is defined as the superposition of the angular aspects in the respective planet constellation. These resonance values exhibit complex temporal dynamics resulting from the planetary revolutions.

In astrology, this "property" of time is often called the "astrological quality (of the moment) of time" (e.g., Niehenke 2002, Hyde 1992:127)[12] or "momentary quality" (e.g., Dean, Mather, & Kelly 1996:48). This "quality of time" can be definitely identified for every point in time on the basis of astronomical facts. Every day on the time line has its particular angular aspect structure between the planetary positions and, with that, its particular possibility of building specific angular aspects to the planets' position (R) at any other point on the time line. Our method of quantifying R allows us to calculate the resonance value of every point on the time scale corresponding to an individual horoscope. This calculation must be done separately for every candidate because of the particular astronomical situation on their date of birth.

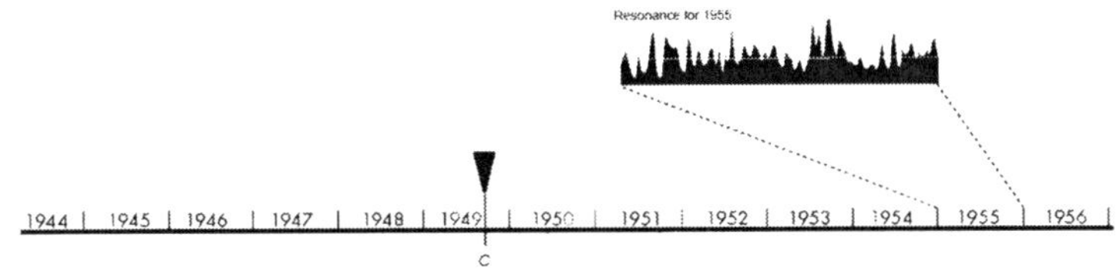

Figure 3. Time scale related to a candidate (C) with the fluctuation of the *R* in the year 1955 shown as an example.

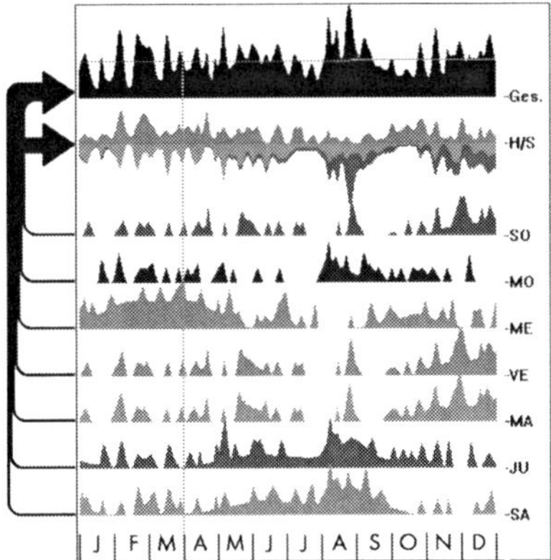

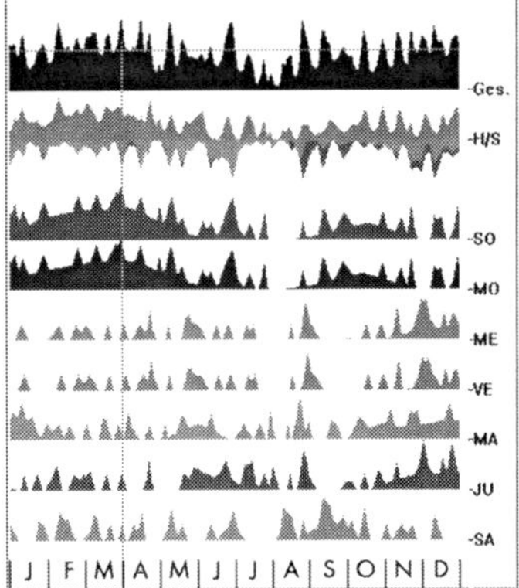

Figure 4. Example diagrams of resonance values from two different candidates A and B.

The *y*-axis shows the degree of the quantified resonance; the *x*-axis shows the time segment of 1955 with the total of 365 days. The first row shows the total values of all planet resonances relating to the individual charts of A and B, respectively. The second row shows the total values separated into dissonant (light grey) aspects, harmonic (medium grey) aspects, and conjunctions (dark grey). The other rows show the resonance values of the single planets (Sun–Saturn). The diagrams clearly show the considerable daily fluctuation of the amount and the intensity of the interaspects. Short frequency fluctuations caused by the fast-moving planets (Sun to Mars) are superposed by long frequency fluctuations caused by Jupiter, Saturn, and the Trans-Saturnal planets. It also becomes apparent how different the resonance values are for the two candidates A and B, depending on the birth charts. The vertical line in the diagram marks a single day at the end of March 1955 as an example to show the considerable individual differences of the resonance values.

The individual assessment of chance expectancy is necessary due to the inhomogeneity of the time axis, that is due to the permanently changing astronomical constellations of every point in time (date of birth). The values of individual chance expectancy can be assessed by calculating the R of a candidate's chart with the chart of every day on the time axis over a long period, and then by averaging the single resonances (see Figures 3–5, see Mathematical Appendix). Choosing a plus/minus 15-year period before and after a candidate's date of birth, the resonance value converges to a limit. This limit value can be interpreted as the individual value of chance expectancy (reference value) of the R accounting for the partners' dates of birth on a case-by-case basis because the dates of birth of all possible partners within this time period of 30 years have been taken into account with this procedure (see Figure 9).

However, there are further (socio-) psychological factors that have to be taken into consideration. People more frequently enter into loving relationships or relationships as friends when there is a relatively small difference between the ages of the partners. This is immediately evident: In school, in professional education, and in recreational activities, people come predominantly into contact with people of the same peer group. Indeed, there may be particular preferences in individual cases—for instance, being explicitly attracted to older people (see Figure 6). Thus, there are substantial uncertainties in estimating the "true" individual value of chance expectancy due to these factors which are difficult to manage. Therefore we developed a procedure in order to take such individual preferences and also general tendencies (peer group effects) into account: We considered the time segments around the dates of birth of the partners (current as well as previous) to assess the individual reference values by a randomized "blurring" of the partners' dates of birth (Monte Carlo method, see Mathematical Appendix) in accordance with a Gaussian distribution (computer simulations). With this method we were able to generate individual reference values that are not based on theoretical speculations about the chance expectancy of the choice of partners combined with the necessity of considering a wide variety of sociological and socio-demographical aspects. Instead, we gained this on a case-by-case basis of the possible birthdays of partners related to their chosen assemblage (the distribution of the actually indicated partners' dates of birth along the time axis).

One of the great benefits of this approach lies in the fact that all particular constellations in the chart of the candidate as well as all constellations of the relevant time periods are automatically taken into account because the selection of these time periods is prespecified by the birth dates of the particular candidates and of their actual partners.

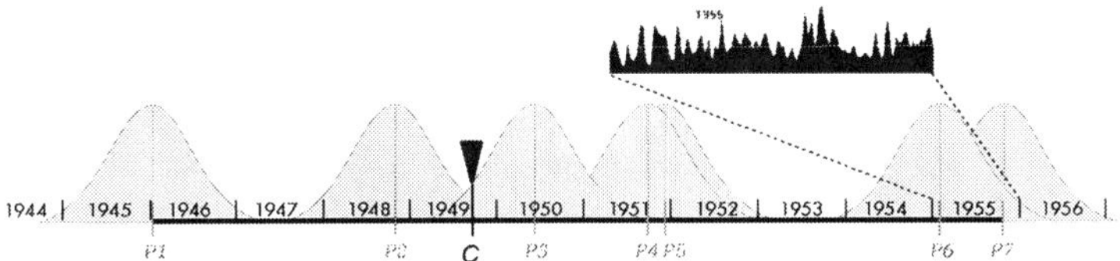

Figure 5. C is a candidate's date of birth, and P1–P7 are the partners' dates of birth.
The *x*-axis, as a time axis, contains the individually calculated resonance curve of C over a long period of time (the enlarged black time segment demonstrates the course of the resonance curve). The *y*-axis shows the frequency distribution of the tests with single time points (dates of birth) within the simulation process statistics (grey curves).

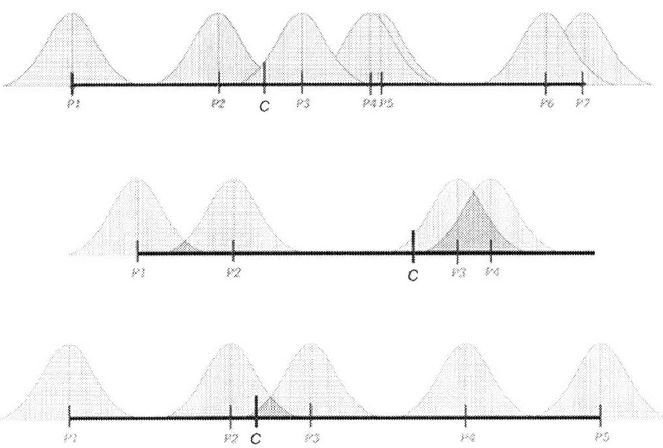

Figure 6. Different distributions of partners' dates of birth with three different candidates.
The first candidate has seven, the second four, and the third five partners.

However, there are a few parameters that have to be determined partly on the basis of theoretical considerations and partly on the basis of empirical testing. One unproblematic point concerns the number of computations used to generate the reference values. We chose K = 5,000 computations (= 5,000 potential partner dates of birth per candidate distributed around the dates of birth of the actually chosen partners of the respective candidate) due to the observation that the changes in the results are negligible above this point.

The choice of the magnitude of the time periods around the partners' dates of birth ((modified) Gaussian distribution) which was used for the computer simulations was much more sophisticated, mainly because of the periods of revolution of the slower planets. On the one hand, the lower limits should not be too narrow because the slower planets should have the possibility of having an effect. On the other hand, the upper limits should not be wide because this would level off the particular patterns of partners. Another consideration concerns the question of whether it is reasonable to vary the time period corresponding to the age differences between the candidate and the particular partner. This may be plausible from an external point of view—from an astrological point of view it is not necessarily the case. Taking these considerations into account, we decided to use two different simulation models: one with a constant variance of two years around every partner's date of birth (SM 1—referring more to astrological reasoning) (see Figure 7), and one with a variable variance corresponding to the age differences, with a minimum time period of one year and a maximum time period of five years, and a linear gradient of 0.25 (SM 2—referring more to external criteria such as socio-demographical reflections) (see Figure 8).[13] For comparative purposes, we also performed computations with a constant time period of plus/minus 15 years (equally distributed) around the candidate's date of birth without accounting for the partner's date of birth (SM 3) (see Figure 9).

Data Collection

As mentioned above, the data were collected in two phases. The first sample was collected using paper questionnaires distributed among some *Deutscher Astrologenverband* (DAV) training centers and among our circles of acquaintances. We received 137 candidate questionnaires in response, with a total of 1186 dates of birth of partners (mean = 8.65 per candidate). 98 candidates were female, 36 were male, and 3 were not stated. 610 relationships were characterized as particularly intensive.

The second data sample for the conceptual replication was gained via the Internet. 227 participants responded (female = 168, male = 59) and

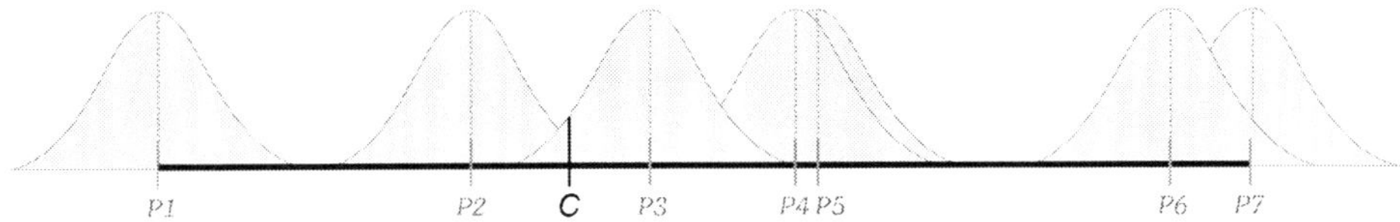

Figure 7. **SM 1: Normally distributed, with a constant standard deviation of two years around every partner's date of birth (P1–P7).**
C = candidate's date of birth.

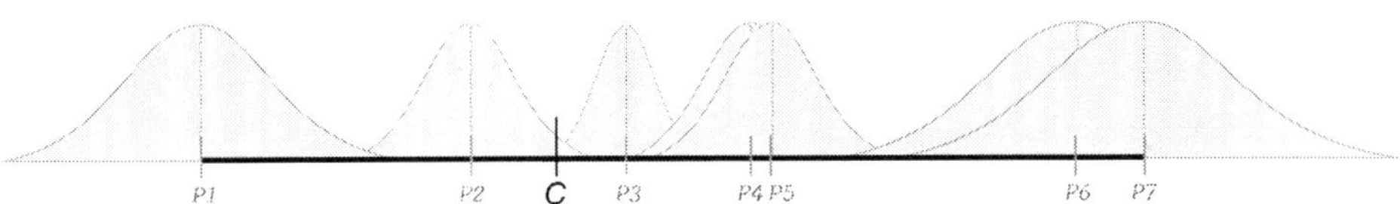

Figure 8. **SM 2: Normally distributed, with a variable standard deviation depending on age differences approximately proportional to age difference (see Mathematical Appendix).**

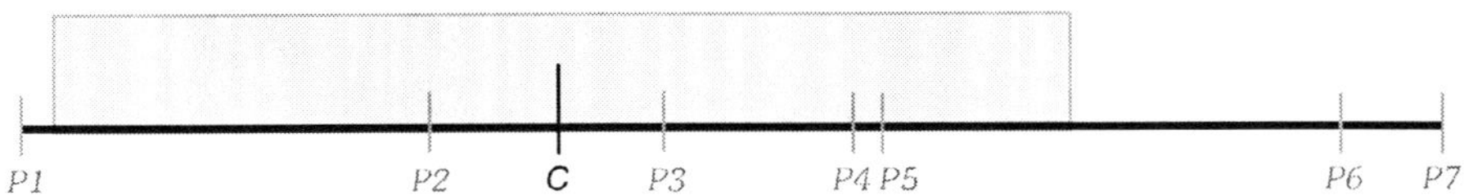

Figure 9. **SM 3: Uniform distribution with a constant time period of plus/minus 15 years around the candidate's date of birth, without accounting for the partners' dates of birth (SM 3).**

provided a total of 1649 datasets of partners (mean = 7.26 per candidate). A total of 610 relationships were characterized as particularly intensive. The statement concerning the knowledge of one's own sun sign was negated by 27 individuals, and should be interpreted as knowing little regarding astrology. 188 individuals indicated that they knew the zodiacal sign of their ascendant. 164 candidates mentioned that they have knowledge of the Moon position in their birth chart. Table 1 comparatively lists the data from both samples.

TABLE 1
Characteristics of the Two Samples

	Sample 1	Sample 2
Number of candidates	137	227
Female	98 (71.5%)	168 (74%)
Male	39 (28.5%)	59 (26%)
Mean age	41.6	39.1
Total amount of partners	1186	1649
Mean number of partners per candidate	8.65	7.26
Relationships characterized as particularly intensive	613 (52%)	917 (56%)
"I know the properties of my sun sign well"	Not surveyed	200 (88%)
"I know my ascendant"	Not surveyed	188 (83%)
"I know the zodiacal position of the Moon at the time of my birth"	Not surveyed	164 (72%)
"If I make friends with someone I ask about the person's Sun sign"	Not surveyed	122 (54%)
"Is the Sun sign an important factor to you regarding the choice of your friends?" *("Yes" responses)*	9 (7%)	Not surveyed
"Have you already read a book on astrology?" *("Yes" responses)*	102 (74%)	Not surveyed

Results

The analysis of the first sample provided some promising results, but also contained elements that did not support the hypotheses. We found a significant deviation of chance expectancy in the expected direction regarding the main hypothesis (H 1, computed with Simulation Model 1), that is there were more or more accurate interaspects between the chart of the candidate and the charts of their friends and people they were in a long-term relationship with than could be expected randomly (on the basis of individual reference values). Regarding the fourth hypothesis, the deviation of chance expectancy is of the same order as H 1, but just under the significance level of 5%, that is we found more resonance than could be expected randomly between the "most resonant" planet of the candidate and all planets of the corresponding partners. The results for the other hypotheses showed the same tendencies, but lower z-values (Table 2).

Most surprising were the results for our sub-hypotheses on the intensity

TABLE 2
Results for the Five Hypotheses
with the Three Different Simulation Models

Sample 1	SM 1	SM 2	SM 3
H 1: Resonance—total	1.7*	1.48	0.8
H 2: Resonance (prespecified)	1.48	1.46	1.03
H 3: Resonance to Saturn (partners)	1.14	0.75	−0.23
H 4: Most resonant planet (candidate)	1.56	1.44	0.99
H 5: Most resonant planet (partners)	1.38	1.21	0.77

$N = 137$ candidates; t-test; z-values; * = Significant at the 5% level. The data for SM 3 are provided for comparative reasons. The considerable difference between the results computed for SM 1/SM2 as opposed to SM 3 clearly shows the relevance of our specific method which takes the individual preferences, as well as socio-demographic factors, into account.

of the relationships. We assumed that the resonance should increase correspondingly with intensity. This was clearly not the case. In contrast, we found highly significant differences in the unexpected direction (Table 3).

These unexpected results from the first sample prompted us to do a conceptual replication of our study with an exploratory part to gain more detailed information on the specific nature of the relationships. We hoped that such additional information could explain these inexplicable results re-

TABLE 3
Differences between the Resonance of "Normal" Relationships
and Relationships Characterized as "Intensive"

Sample 1	SM 1		SM 2	
	Normal N = 466	Intensive N = 613	Normal	Intensive
H 1: Resonance (total)	2.82**	−0.3	2.72**	−0.50
H 2: Resonance (prespecified)	2.8**	−0.49	2.8**	−0.46
H 3: Resonance to Saturn (partners)	2.09*	−0.54	1.84*	−0.77
H 4: Most resonant planet (candidate)	2.1*	−1.16	2.04*	−1.25
H 5: Most resonant planet (partners)	2.50**	0.38	2.49**	0.24

t-test; z-values; * = significant on the 5% Level, ** = significant on the 1% level

garding the sub-hypotheses, and additionally strengthen our main findings. Surprisingly, we received a much more unexpected picture with the new results: The main effects seen in the first sample disappeared, and that of the sub-hypotheses stood partly in contrast to the results of the first sample (Table 4):

TABLE 4
Results of Sample 2 for the Five Hypotheses
with the Two Different Simulation Models

Sample 2	SM 1			SM 2		
	All	Normal	Intensive	All	Normal	Intensive
H 1: Resonance (total)	0.27	−1.37	1.2	0.05	−1.56	1.04
H 2: Resonance (prespecified)	−0.22	−1.4	0.56	−0.26	−1.46	0.5
H 3: Resonance to Saturn (partners)	0.77	0.32	0.64	0.34	−0.05	0.37
H 4: Most resonant planet (candidate)	0.67	0.05	1.44	0.49	−0.08	1.35
H 5: Most resonant planet (partners)	1.12	−0.99	2.21*	0.91	−1.21	2.11*

t-test; *z*-values; * = significant on the 5% Level

Confronted with these contradictory findings, we speculated about the possibility that the second sample collected using the Internet differs considerably from the first sample. However, a closer look at some detailed results led us to the assumption that the significant correlations in the first sample most likely have to be ascribed to chance. Due to the fact that both samples were large enough, we decided to replicate the statistics with a split-half method. As can be seen in Table 5, the results of the halved samples show significant differences.

TABLE 5
Results of the Split-Half Evaluation of Both Samples

	Sample 1		Sample 2	
	A	B	A	B
H 1: Resonance (total)	1.75	0.64	0.74	−0.44
H 2: Resonance (prespecified)	1.66	0.43	0.29	−0.64
H 3: Resonance to Saturn (partners)	1.71	−0.1	0.3	0.81
H 4: Most resonant planet (candidate)	2.12	−0.02	−0.12	1.11
H 5: Most resonant planet (partners)	1.18	0.73	0.85	0.61

SM 1; *t*-test; *z*-values

With these results our null hypotheses are confirmed: There is no more resonance than could be expected randomly between the charts of friends and people in a long-term relationship, respectively. This applies for "normal" relationships as well as for "intense" relationships.

Discussion

The findings differ from the expectations at the beginning of the research project. We were aware that if the astrological correspondence hypothesis is true we could not expect to see substantial effects: If everything is considered together, and, moreover, if important astrological factors for the analysis of astrological synastries are missing (e.g., the Moon positions), the result is inevitably a slightly faded image (cf. Koch 2002:131). We nevertheless assumed that an effect would be apparent by using a large sample, or that at least a tendency would become visible which one could make distinct by adjusted methodological means. This, however, was not the case. Rather the contrary: The results of the split-half tests provide a picture that can be best described as fluctuation by chance. Further explorative analyses of subgroups (e.g., number of partners, sex of candidates and partners) did not give any indications as to systematic significant deviations from chance. Therefore the obvious conclusion seems to be that there exists no astrological effect. So if one does not want to abandon the general hypothesis of a significant correspondence of the choice of friends and long-term relationships and of the interaspects of their charts despite the findings, it is only justifiable under the assumption that the correspondence is much more concealed than we postulated in our hypotheses. Maybe we underestimated the impact of individual differences in assessing relationships. Another possibility is that the wide range of numbers of partners per candidate (from 1 to 22 partners in Sample 1; in Sample 2 the technically determined upper limit of partners was 12) led to an as-yet-uncontrollable bias.[14] To verify such possibilities one has to perform increasingly subtle tests with the data on an exploratory basis, for example by taking the kind of relationship (love relationship, business relationship, etc.) systematically into account (as a variable), but also to differentiate between hard and soft aspects, to include minor aspects, to weight the various astrological factors, etc. This work has yet to be done. The fact that the *t*-statistics assume relatively high values could be seen as an indication that there is some systematic factor at work. On the other hand, some large test values are to be expected due to multiple testing also when there is no effect.[15] The issue would require careful consideration in a study claiming positive evidence which is not the case here.

We have not found any significant impact of astrological knowledge on the participants in our results as we hypothesized because of the complexity of the methodological design. If there had been attempts by participants to influence the results by preselecting the provided data in order to support the astrological hypothesis (based on hypotheses about the presumed aim of the astrological investigation), these attempts did not lead to a systematic uni-directional bias in the data. However, the impact of such strategies on the results is difficult to estimate. Maybe it represents such a hidden systematic factor.

Our investigation of a possible "as above, so below" association using a high complexity of astrological factors failed to provide corresponding evidence. Future studies on a similar topic might pursue the following two directions: One option would be to further increase the complexity by incorporating additional relevant astrological factors such as the Moon position, the ascendant, the descendent, and the 7th house (factors which are regarded as astrologically important for relationships). That would support the astrological hypothesis that the astrological factors are multifunctional and ambiguous to a certain extent: Different factors are said to have similar meanings or impacts, and one and the same factor has different impacts on different levels and in various contexts. With our methodological tool, the inclusion of further factors would not cause any difficulties, but the gathering of data is much more time-consuming. The second direction would be to place an emphasis on differentiating the relationships, enabling a more accurate characterization. This would imply a return to the method of relating particular aspects of the total complex of a birth chart to particular characteristics of relationships. Thus, a lot of explorative work has to be done.

Conclusion

Against the background of these results which are negative from the astrological perspective, the main—and absolutely positive—conclusion concerns, on the one hand, a better understanding of the complexity of the astronomical situation with regard to astrological investigation. The substantial differences between the first two simulation models related to individual patterns of choice of the partners, and the third model without that reference (see Table 2) clearly points to this. It results in empirical evidence regarding the necessity to consider the inhomogeneity of the time scale (in an astrological sense). This was not considered sufficiently in previous studies that had a related approach (e.g., Ruis 1994, 1994/1995).

On the other hand, we developed a new methodological approach for astrological research. The research tool created for our study is applicable for many other research issues: One could, for instance, reassess the astrological heredity hypothesis made by Gauquelin (1978:177–186) (see also Brady 2002) with our method of quantifying the astrological resonances of the charts of parents and their children. The horoscopes of married couples could additionally be investigated regarding particular qualities and dynamics of the relationship and corresponding resonances of particular planets. The crucial differences to traditional approaches lie: (1) in the individual assessment of the chance expectancy of particular astrological constellations on a case-by-case basis, and (2) in the consideration of individual, social–psychological, and demographic influences by taking the actual (i.e. not hypothesized on the basis of statistical values) patterns of relationships with their individual structures of time distances into account. With this, we created a new approach to performing quantitative astrological studies without the necessity of making use of major speculations on either theoretically derived values of chance expectancy or on socio–demographical influences.

Acknowledgment

We thank Werner Ehm for his invaluable help in operationalizing our research question and for developing suitable statistical procedures for the analysis of the data. He also added the Mathematical Appendix section providing mathematical and statistical details.

Notes

[1] The main problem of the latter lies in the incorrect assumption that an astrological sign of the zodiac (which stands for a distinct characteristic of every segment of 30° with an abrupt change of its property at the transition points) is a concrete fact, comparable to facts such as biological gender, being married, or holding a high school diploma. For scientific purposes one has to deal with zodiacal signs as if they are mere human constructions of a highly hypothetical nature, and one has to account for this fact methodologically for example by testing the correlations found with the zodiacal division of the hypothetical circle against other circle graduations. Niehenke (1998) mentions this point in an exemplary way in his critique of the book *Die Akte Astrologie* by Sachs (1997) (translated and published in English under the title *The Astrology File* (Sachs 1998)).

[2] One of the important approaches in astrological research that provided the most convincing results with regard to the astrological hypothesis is the work done by the Gauquelins (1978, 1983, 1988b) which was followed by comprehensive re-analyses and complementary work/comments by Ertel and others (e.g., Ertel & Irving 1996, Ertel 2011). This tradition of research is only of minor relevance for our approach because the Gauquelin-type results do not fit into the usual astrological system of interpretation—at least not at first sight. Hence, Gauquelin proposed a new form of astrology which he called "Neo-Astrology" (1991).

[3] A prominent example is the famous German nuclear physicist and philosopher Carl Friedrich von Weizsäcker, who got involved with the complex analysis of horoscopes in the 1940s, did a fair amount of his own practical work during this time, and who still advanced almost forty years later his opinion regarding his personal experience of evidence of astrological analysis—notwithstanding his critical attitude toward astrology as a social practice of counseling:

> I, as a physicist, have no rhyme or reason at all up to today what should be the case if astrology were empirically true. On the other hand, I got the impression—simply by my preoccupation with it—that there is something in it. (von Weizsäcker, quoted in Niehenke 1987:22, translation by the authors)

More references could be given but most of them are only private statements because scientists are reluctant to advocate astrology publicly. However, an academic education in natural sciences is no guarantee to be able to make qualified judgments outside one's studied discipline, of course.

[4] One could interpret this as a kind of proto-statistical procedure.

[5] A lot of typical arguments made by astrologers as well as critical researchers concerning the problems of astrological research are compiled in Phillipson (2000:passim).

[6] Of course, we are aware that from an astrologer's perspective the image of the partner and the motivation toward a relationship cannot be reduced to the interaspects of the two corresponding charts from an astrological perspective. However, it should be broadly accepted by astrologers that they make a substantial contribution to the choice of partners.

[7] This also concerns the above-mentioned sun-sign effect which has been found by Eysenck and Nias (1982) because our study is not based on a response to personality traits, but on the existence of a relationship. However, with regard to the assessment of the intensity of the relationship, the possibility of an influence of the sun-sign effect should be considered.

[8] The exclusion of the trans-Saturn planets was made to avoid long-term effects that influence a generational cohort. The exclusion of the Moon had to be made due to the fact that the time of birth of the partners was lacking. This was very unfortunate because the Moon plays an important role in relationships from an astrological perspective, but this could not be avoided for reasons relating to the practical execution of the investigation.

[9] For exploratory reasons, we additionally computed the other partner planets.

[10] Ashmun (1984) pointed out this problem in her critique of Carl Jung's astrological experiment and expressed her hope that computer technology would help in the future to solve this laborious task. A quarter of a century

later, the latter point is not a problem anymore. Gauquelin's work provoked a lot of critical analysis on the topic of expectation values of astronomical/ astrological frequencies. However, most of it does not consider complex astrological aspect constellations. In most of Gauquelin's works he focused on the expectation values of one planet in different horoscope sectors.

[11] Ertel applied this method in his investigation into the relation of planetary aspects with human birth dates (Ertel 1988). It was a suitable strategy for his research dealing with single aspects related to one date (date of birth).

[12] An astrological journal edited by the Österreichische Astrologische Gesellschaft (Austrian Astrological Society) is entitled *Qualität der Zeit* (*Quality of Time*). In English-speaking countries this expression does not seem to be as common as in German-speaking countries.

[13] From a socio-demographic viewpoint, the assessment of an age difference in partners' dates of birth is strongly dependent on the age of the candidate: If the candidate is, for example, 16 years old, one of his partners is 17, and the other 22 years old, the age difference of five years between the partners is regarded as more significant than if the one friend is 43 and the second 48 years old. As a result, an increasing standard deviation around the partners' dates of birth corresponding to the age differences from the candidate's date of birth is indicated (SM 2). From an astrological perspective, the variance around the partners' dates of birth has to be kept constant, independent of the age differences with the candidate's birthday because of the inhomogeneity of the above-mentioned time scale (SM 1).

[14] We considered this partly by an exploratory use of different basic units. Instead of the originally chosen groupwise evaluation (one candidate together with her partners as a basic unit—see Mathematical Appendix), we performed a pairwise evaluation (one candidate with one of her partners as a basic unit) with the different kinds of evaluation of the candidate–partner complexes (pairwise vs. groupwise). This difference is relevant because of the different numbers of partners provided by the candidates. In the groupwise mode, the relationships of a candidate with only a few partners are weighted more than those of a candidate with a lot of partners. In contrast, the pairwise mode places more weight on the candidates with a lot of partners. It is nearly impossible to predict the effect of these biases. However, the pairwise mode is not consistent with Hypotheses 4 and 5 because they are made relating to groups of partners. The results of the pairwise evaluation differ slightly but not substantially from the results of the groupwise evaluation.

[15] The extent of the alpha error inflation is difficult to quantify properly because of stochastic dependencies between the test statistics. Some further comments on this point may be found in the Mathematical Appendix.

References

Arroyo, S. (1978). *Astrology, Psychology, and the Four Elements: An Energy Approach to Astrology and Its Use in the Counseling Arts*. Davis, CA: CRCS Publications.

Ashmun, J. (1984). Critique of C. G. Jung's Astrological Experiment. *Astro–Psychologische Probleme, 2*(2), 28–32.

Böer, W., Niehenke, P., & Timm, U. (1986). Lassen sich "Unfäller" astrologisch diagnostizieren? Ein exploratorisches Experiment. *Zeitschrift für Parapsychologie und Grenzgebiete der Psychologie, 28*, 56–65.

Brady, B. (2002). The Australian Parent–Child Astrological Research Project. *Correlation, 20*(2), 4–20.

Davison, R. (1983). *Synastry. Understanding Human Relations through Astrology*. New York: Aurora Press.

Dean, G. (1977). *Recent Advances in Natal Astrology: A Critical Review, 1900–1976*. Analogic.

Dean, G. (1985a). Can astrology predict E and N? 1. Individual factors. *Correlation, 5*(1), 3–17.

Dean, G. (1985b). Can astrology predict E and N? 2. The whole chart. *Correlation, 5*(2), 3–24.

Dean, G. (1986). Can astrology predict E and N? 3. Discussion and further research. *Correlation, 6*(2), 7–52.

Dean, G. (1999). Astrology and human judgement. *Correlation, 17*(2), 24–71.

Dean, G., & Mather, A. (1994). Is the scientific approach relevant to astrology? *Correlation, 13*(1), 11–18.

Dean, G., Mather, A., & Kelly, I. W. (1996). Astrology. In G. Stein, Editor, *Encyclopedia of the Paranormal*, pp. 47–99, Amherst, NY: Prometheus.

Denness, B. (2000). Thieves, victims, and the zodiac. *Correlation, 19*(1), 42–48.

Ertel, S. (1988). Relating planetary aspects to human birth: Improved method yields negative results. *Correlation, 8*(1), 5–21.

Ertel, S. (1998). Astro-Quiz: Can astrologers pick politicians from painters? *Correlation, 17*(1), 3–8.

Ertel, S. (2011). Rückblick (1955–2005) auf die durch Michel Gauquelin entfachte Forschung. In U. Voltmer & R. Stiehle, Editors, *Astrologie & Wissenschaft*, pp. 280–323, Tübingen: Chiron.

Ertel, S., & Irving, K. (1996). *The Tenacious Mars Effect*. London: Urania Trust.

Eysenck, H. (1981). The importance of methodology in astrological research. *Correlation, 1*(1), 11–14.

Eysenck, H. J. (1983a). Happiness in marriage (Part I). *Astro–Psychological Problems, 1*(2), 14–23.

Eysenck, H. J. (1983b). Happiness in marriage (Part II). *Astro–Psychological Problems, 1*(3), 15–19.

Eysenck, H. J., & Nias, D. (1982). *Astrology: Science or Superstition?* London: M. T. Smith.

Gauquelin, M. (1978). *Cosmic Influences on Human Behavior*. New York: ASI Publications.

Gauquelin, M. (1983). *The Truth about Astrology*. Oxford: Blackwell.

Gauquelin, M. (1988a). *Planetary Heredity* (Foreword by Giorgio Piccardi). San Diego: ACS Pub.

Gauquelin, M. (1988b). *Written in the Stars: The Proven Link between Astrology and Destiny*. Wellingborough: Aquarian.

Gauquelin, M. (1991). *Neo-Astrology: A Copernican Revolution*. London: Arkana.

Hyde, M. (1992). *Jung and Astrology*. London: Aquarian/Thorsons.

Jung, C. G., & Main, R. (Eds.) (1997). *Jung on Synchronicity and the Paranormal. Key Readings*. London: Routledge.

Kelly, I. (1979). Astrology and science: A critical examination. *Psychological Reports, 44*, 1231–1240.

Klein, M. (1988). The accuracy of relationship description as a test of astrology. *Correlation, 8*(2), 5–17.

Koch, D. (2002). *Kritik der Astrologischen Vernunft. (2. Überarbeitete und Stark Erweiterte Auflage).* Frankfurt/Main: Verlag der Häretischen Blätter.

Kollerstrom, N., & O'Neill, M. (1992). Invention moments and aspects to Uranus. *Correlation, 11*(2), 11–23.

Kuypers, C. (1984). A study on marriages. *Astro–Psychological Problems, 2*(4), 17–21.

Meyer, M. R. (1976). *The Astrology of Relationship. A Humanistic Approach to the Practice of Synastry.* Garden City, NY: Anchor Press.

Müller, A. (1957). Eine statistische untersuchung astrologischer faktoren bei dauerhaften und geschiedenen ehen. *Zeitschrift für Parapsychologie & Grenzgebiete der Psychologie, 1,* 93–101.

Nanninga, R. (1996/1997). The Astrotest: A tough match for astrologers. *Correlation, 15*(2), 14–20.

Niehenke, P. (1987). *Kritische Astrologie. Zur Erkenntnistheoretischen und Empirisch-Psychologischen Prüfung ihres Anspruchs.* Freiburg im Breisgau: Aurum.

Niehenke, P. (1998). *The Astrology File*: Scientific proof of sun sign effects? *Correlation, 17*(1), 41–44.

Niehenke, P. (2002). *Die Metapher von der Qualität der Zeit. Das Wesen der Zeit und das Wesen des Menschen.* Proceedings of the DAV Congress, Geburt–Tod–Transzendenz (Birth–Death–Transcendence); 28–30 June 2002; Munich. http://www.astrologiezentrum.de/onlinetexte3.html

O'Neill, M. (1986). The moon's nodes in synastry. *Astro–Psychological Problems, 4*(3), 24–32.

O'Neill, M. (1989). The moon's node in marriage. A replication. *NCGR Research Journal,* 21–25.

O'Neill, M. (1990). The moon's nodes in synastry—A replication. *Correlation, 10*(1), 10–18.

O'Neill, M. (1995). Further comments on Jan Ruis's marriage results. *Correlation, 14*(1), 26–29.

Phillipson, G. (2000). *Astrology in the Year Zero.* London: Flare Publications.

Ptolemy, & Ashmand, J. M. (1822). *Ptolemy's Tetrabiblos: Or Quadripartite: Being Four Books of the Influence of the Stars.* Newly translated from the Greek paraphrase of Proclus with a Preface, Explanatory Notes, and an Appendix containing extracts from the *Almagest* of Ptolemy and the whole of his *Centiloquy* together with a short notice of Mr. Ranger's Zodiacal Planisphere and an explanatory plate. London: Davis & Dickson.

Ruis, J. F. (1993/1994). Indication for a role of synastry aspects in a "Gauquelin"—Sample of 2,824 marriages (1). *Correlation, 12*(2), 20–43.

Ruis, J. F. (1994). Indication for a role of synastry aspects in a "Gauquelin"—Sample of 2,824 marriages (2). *Correlation, 13*(1), 3–9.

Ruis, J. F. (1994/1995). Shift control of synastry–effect. *Correlation, 13*(2), 38–39.

Ruis, J. (2007/2008). Statistical analysis of the birth charts of serial killers. *Correlation, 25*(2), 7–44.

Sachs, G. (1997). *Die Akte Astrologie.* München: Goldmann.

Sachs, G. (1998). *The Astrology File.* London: Orion.

Shanks, T. (1983). Research on astrological factors between married couples. *Astro–Psychological Problems, 2*(1), 12–16.

Shanks, T., & Steffert, B. (1984). Planets and happiness in marriage. *Astro–Psychological Problems, 2*(3), 5–9.

Smithers, A., & Cohen, D. (1982). Personality, sun-sign, and planetary position. *Correlation, 2*(2), 12–20.

Steffert, B. (1983). Teilprüfung gegen gesamtprüfung des horoskops. *Astro–Psychologische Probleme, 1*(2), 20–25.

Van de Moortel, K. (1998). Love at first sight . . . a study into astrology and attraction. *Correlation, 17*(1), 50–53.

Mathematical Appendix

Resonances

The basic units of the statistical analysis consist of a candidate together with her partners. Initially, each such unit is analyzed on its own. The combined results of the individual analyses then allow statements about the database as a whole.

For a fixed unit, let CT denote the birth time of the candidate, and $PT_1, \ldots, PT_N$ the birth times of her partners, time being measured according to the Julian calendar. If only the birthday is known, the birth time is set to noontime. Given any two planets p_1, p_2 and any two time points t_1, t_2, let $\alpha = \alpha\,(p_1, t_1; p_2, t_2)$ denote the angle between the positions of planet p_1 at time t_1 and planet p_2 at time t_2 as seen from the earth. The resonance of such a constellation depends only on the angle α (taken modulo 180°). It is given by the resonance function $\rho(\alpha)$ which has a tent-like deflection at the astrologically meaningful angles of 0, 60, 90, 120, and 180 degrees, and is zero if α differs from all these by 5° or more.

The planets entering the analysis differ across hypotheses and also between candidate and partners. For a given hypothesis H let CP and PP denote the set of planets that are relevant to the candidate and her partners, respectively. The basic resonance statistic for the given unit and the hypotheses H_1, H_2, H_3 then is the average resonance evaluated at the respective birth times,

$$R_H = \frac{1}{N n_{cp} n_{pp}} \sum\nolimits_{p \in CP,\, p' \in PP} \sum\nolimits_{i=1}^{N} \mathsf{r}\,(\mathsf{a}\,(p, CT; p', PT_i)) \quad (1)$$

Here n_{cp}, n_{pp} denote the numbers of candidate and partner planets, respectively. For $H3$, for example PP consists of Saturn only, so $n_{pp} = 1$. In Hypothesis 4 and Hypothesis 5, one of the two averages across planets is replaced by a maximum, for example

$$R_{H_4} = \max\nolimits_{p \in CP} \frac{1}{N n_{pp}} \sum\nolimits_{p' \in PP} \sum\nolimits_{i=1}^{N} \mathsf{r}\,(\mathsf{a}\,(p, CT; p', PT_i)) \quad (2)$$

$$R_{H_5} = \max\nolimits_{p' \in PP} \frac{1}{N n_{cp}} \sum\nolimits_{p \in CP} \sum\nolimits_{i=1}^{N} \mathsf{r}\,(\mathsf{a}\,(p, CT; p', PT_i)) \quad (3)$$

Reference Distributions and p-Values for a Single Unit

The above resonance statistics are evaluated at the actual ("observed") birth times of the candidate and her partners; we indicate this by writing RH^{obs}. Our reference distribution for this value is obtained by randomizing the partner birth times as follows. (The candidate birth time is kept fixed.)

Let $R_H(T_1, \ldots, T_N)$ denote the resonance statistic obtained when each partner birth time PT_i is replaced by some other time T_i. (Note that then $RH(PT_1, \ldots, PT_N) = R_H^{obs}$, according to our conventions.) Suppose the times $T_1, \ldots, T_N$ are statistically independent random variables with distributions $F_1, \ldots, F_N$ selected according to one of the three sampling schemes. (More details on this are given below.) When plugged into the resonance statistic, $R_H(T_1, \ldots, T_N)$ itself becomes a random variable. *Our reference, or null-distribution, D_0, then is defined as the distribution of this random variable, $D_0 = distribution\ of\ R_H(T_1, \ldots, T_N)$.* It represents the respective null-hypothesis in technical terms, and makes precise what we understand under chance expectation (for a given sampling model).

The distribution D_0 is very complicated and cannot be calculated explicitly. However, an approximation to D_0 is readily obtained by Monte Carlo simulation. Using a (pseudo-) random number generator one generates a large number K of partner birth time arrays $(T_1^{(k)}, \ldots, T_N^{(k)})$ according to the respective sampling model (here $K = 5000$). This gives a distribution D_0^* of K simulated resonance values $R_H(T_1^{(k)}, \ldots, T_N^{(k)})$ which we take as a substitute for the unaccessible "exact" null-distribution D_0.

The p-value, p_H, for testing the null-hypothesis H on the given unit is given by the upper tail of D_0^* beyond R_H^{obs}, i.e. by the fraction of simulated resonance values that exceed the observed value R_H^{obs}. As usual, a small p-value is speaking against the null-hypothesis H.

For an intuitive explanation, note that evidently from the wiggly time course of the resonance functions (see Figures 3 to 5), R_H^{obs} will be (relatively) large if the partner birth times happen to fall on peaks of the resonance functions. These peak times form a highly fragmented, finely structured subset of the time axis. Under the sampling models, alternative partner birth times are sampled much more homogeneously in time, thus reflecting the null-hyothesis that entering a relationship does *not* depend on temporally fine-structured astrological synastries.

Instead of p-values one also may consider z-values, defined as $z = (R_H^{obs} - \mu)\,/\,\sigma$ where μ and σ^2 denote the mean and the variance of the reference distribution D_0^*, respectively. It should be noted, however, that referring such a z-value to the standard normal distribution is meaningful only if D_0^* itself is close to the latter.

Let us yet point out a noteworthy consequence of the above: In contrast with the usual ensemble-based methods, our setting allows us to test astrological hypotheses at the level of a single candidate–partners unit. The aggregation of the individual test results is discussed next.

Overall Statistical Analyses

Everything thus far referred to a fixed candidate–partners unit. Overall significance across units was assessed as follows. Under H, the individual z-values z_j for the single units, though perhaps not normally distributed (see above), have mean zero and variance 1. By the central limit theorem, if the number J of units is not too small, the average z-value times $\sqrt{J}$, that is $\zeta := J^{-1/2} \Sigma z_j$, is approximately standard normally distributed, hence may be used as an (approximate) z-value. Slightly more cautiously, we based our significance tests on the t-statistic $\tau = \zeta / s$ where s is the empirical standard deviation of the $z_j s$.

Of course, since quite a number of tests have been carried out, there is a multiple testing problem causing alpha error inflation. A Bonferroni correction could rectify this, but would tend to be overly pessimistic due to positive correlation of the test statistics. After all, clear support has not been found for astrological synastry, so we took a more liberal stance and reported individual test results without alpha correction. Denying the exploratory character of our study was not intended thereby. (The Mathematical Appendix writer's opinion is that the meaning of "significant" test results is often misrepresented, overdoing grossly what can be obtained from statistical analyses—which is not meant to say they are useless.)

Sampling Models

The sampling models determine the distribution G according to which alternative "surrogate" birth times are randomly drawn and substituted for the respective partner birth time PT.

SM1: G is the normal distribution with mean $\mu = PT$ and constant standard deviation $sd = 2$ (years).

SM2: G is the normal distribution with mean $\mu = PT$ and standard deviation sd depending on the (absolute) age difference $ad = |PT - CT|$ (years) to the candidate as follows: Starting from $sd = 1$ at $ad = 0$, sd grows linearly in ad with slope 0.25 until it reaches $sd = 5$ (at $ad = 16$), from where it remains constant.

SM3: G is the uniform distribution centered at the candidate's birth time CT with half-width 15 years.

Evidently, SM1 and SM2 roughly maintain the unit's particular birth time configuration whereas SM3 disregards it completely.

Stepwise Shifting of Birth Dates on the Time Scale

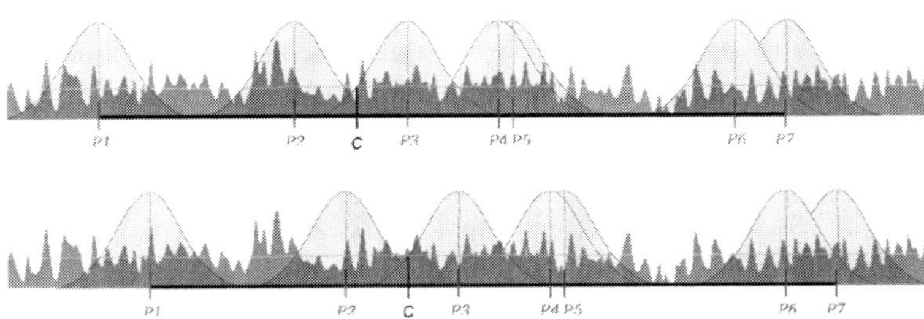

Figure 10. Method of time-shift (suggested by Ertel).
The date of birth of the candidate and all of the partners' dates of birth stepwise shifted on the time scale. This method follows the idea that the actually-found constellation is a "perfect" constellation. According to this, the shifting of the date of birth should lead to a lesser degree of R—comparable with the astrological idea of decreasing effect of an aspect corresponding with its increasing inaccuracy (orbs). This idea is wrong because it is based on the idea of a homogenous time scale. This is not the case (see grey fluctuation of R in the background for illustration).

Journal of Scientific Exploration, Vol. 26, No. 4, pp. 855–865 2012   0892-3310/12

Notes on Early Interpretations of Mediumship

CARLOS S. ALVARADO

Atlantic University, 215 67th Street, Virginia Beach, VA 23451
carlos.alvarado@atlanticuniv.edu

MICHAEL NAHM

Ida-Kerkovius-Strasse 9, 79100 Freiburg, Germany
michaelnahm@web.de

ANDREAS SOMMER

UCL Centre for the History of Psychological Disciplines, University College London, Gower Street, London WC1E 6BT
a.sommer@ucl.ac.uk

Submitted 2/2/10, Accepted 3/7/12

Abstract—The purpose of this note is to dispel the notion that ideas of human agency to account for the veridical mental phenomena of mediums began with persons associated with the Society for Psychical Research (SPR) in England, or with certain later individuals. In fact, the appearance of these ideas preceded the founding of the Society in 1882. Examples of earlier writers who discussed these ideas include Carl Gustav Carus, Edward W. Cox, Justinus Kerner, Asa Mahan, André-Saturnin Morin, Maximilian Perty, B. W. Richmond, and Edward C. Rogers. In contrast to the speculation by later SPR authors and others, the concepts that appeared in the old literature often involved belief in physical forces.

Keywords: Mediumship—human agency—psychic force—spiritualism—
super ESP—super psi

Introduction

The study of the possibility of survival of death is a traditional area of parapsychology, and one that has influenced the field in many ways (Alvarado 2003). Within survival research much attention has been paid to the veridical statements of mental mediums delivered through such means as automatic writing or trance speaking. Among the information that could be corroborated in reports on the mediums of the past were specific names and information about relationships and events relevant to the life of the purported spirit communicator (e.g., Hodgson 1892, Radclyffe-Hall & Troubridge 1919).

Provided conventional explanations such as sensory cues and fraud have been ruled out or considered unlikely, much of this material has been interpreted by some as evidence of discarnate agency. However, as has been discussed in modern writings about mediumship, the psychic abilities of the mediums have been proposed as a frequent counter-explanation for discarnate agency (e.g., Braude 2003, Gauld 1982, Hart 1959, Sudduth 2009). Gauld (1982) stated that the assumption here is that mediums "obtain all their information by telepathically tapping the memory-stores of living persons, clairvoyantly scanning archives, etc." (Gauld 1982:127). The purpose of the present note is to provide the reader with examples of early ideas of this sort, generally neglected in the modern literature.

Recently one author argued that "the idea of explaining mediumistic communications in terms of extraordinary living agent psi originated in the late 19th century among members of the British and American societies of psychical research" (Sudduth 2009:190). Another recent author singled out Frank Podmore (1856–1910), an active member of the Society for Psychical Research (SPR) and a well-known writer, as someone who had developed this concept (Fontana 2005:104). Similarly, authors of overviews of survival of death in psychical research have not mentioned early ideas of this sort (e.g., Almeder 1992, Braude 2003, Fontana 2005, Gauld 1982). Although work conducted by SPR researchers and others was important in the dissemination of the concept of living agency in mediumship (e.g., Leaf 1890), and although their speculations may have been more detailed, the idea is older than that. In fact, it cropped up almost immediately after the rise of Spiritualism. Our review will present examples from the early mesmeric and spiritualistic literature (see also Crabtree 1993, Podmore 1902).

Some Authors Writing on Living Agency and Mediumship Previously to the Foundation of the SPR

The notion that the psychic powers of the living played a role in mediumistic communications attributed to the dead was present in the mesmeric movement. Writing in the journal *Zoist*, English clergyman the Reverend George Sandby (1799–1881) reduced the revelations some mesmeric subjects claimed to come from discarnate spirits to the "old mesmeric principles of thought-reading and clairvoyance" (Sandby 1850:423).[1] In France, lawyer André-Saturnin Morin (1807–1888) disagreed with the belief that mediumistic communications could be explained only by the action of spirits of the dead. In an account of a specific phenomenon, he wrote: "To imitate writings he has never seen . . . the medium reads the thought of the consultant and sees the name of the evoked deceased [person] and the writing at the same time" (Morin 1858:96). In a later publication Morin (1860:464) felt that the information conveyed by mediums was related to sources such as writings or books located in other places accessed through their lucidity (clairvoyance). Similarly, magnetizer Charles Lafontaine (1803–1892) argued that mediums could perceive the "unconscious thoughts of the persons present" (Lafontaine 1866:116).

Discussions of living agency have also been put forward in the early literature of spiritualism. Often, these authors referred to concepts based or derived from mesmerism and entailing the concept of vital, nervous, or psychic forces that were intimately related to the body of the medium. The projection of such a force was assumed by opponents of discarnate agency to be either an automatic process or one guided unconsciously by the medium. Most of these speculations referred to such physical phenomena as raps, table movements, and movement of other objects.[2] An early review of some of these publications in the English and French literature was presented by Alexander Aksákow (1832–1903) (Aksákow 1886).[3] Many such examples existed (e.g., Oldfield 1852, De Gasparin 1854, Dods 1854, Mahan 1855, Samson 1860, Chevillard 1869, Cox 1879). Some of these discussions included the concept of Od, which became popular in the spiritualistic literature at the time.[4] An example was English spiritualist Thomas P. Barkas' (1862) summary of a popular explanation for the phenomena of spiritualism. The explanation rested on the assumption that the phenomena were produced by human agency, among which were

> the automatic action of the cerebral, spinal, and sympathetic nerve centers of mediums, acting through the agency of and using an imponderable and all-pervading fluid designated by Reichenbach, ODYLE. (Barkas 1862:15).

Traverse Oldfield (1852)—a pseudonym used by American minister George W. Samson (1819–1896)—argued that the

> communications of the clairvoyant and of the spiritual medium, as to facts that can be tested, have been only the knowledge, remembered or forgotten, and the surmise, right or wrong, of the person consulting. (Oldfield 1852:44)

He suggested that the explanation was "a united current of two persons' nervous influence" (Oldfield 1852:44) in which the thought of one person passed to the other, and then was expressed through rappings or automatic writing. Such process, Oldfield wrote, is no more mysterious than that, by a connection of electric conductors, and by an excitement of the electric principle, I can control the electric influence of a series of electric conductors, reaching from New Orleans to Boston, and have my thought rapped or written out a thousand miles from the point where I exert the energy. (Oldfield 1852:44)

Physician B. W. Richmond believed that mental phenomena were caused by "mind acting on mind" (Brittan & Richmond 1853:149). The medium, in a magnetic state, was in rapport with another living mind. Phenomena such as automatic writing "follow the law of *mental reflection—*and the *reflected image* often corresponds to other minds than the medium . . ." (Brittan & Richmond 1853:186). He believed that the mind of the medium could not realize that "she is being impressed by a book, a letter, or some human mind" (Brittan & Richmond 1853:204). Furthermore, Richmond asserted that in cases in which a medium produced a facsimile of a deceased person's signature that was unknown to her, the medium, being in rapport with a person familiar with the signature, "copied" it from that person's memory (Brittan & Richmond 1853:204). Another author, and one cited by many at the time, was the American writer Edward C. Rogers (1853). He believed that mediumistic physical phenomena were caused by human nervous forces related to the medium's brain centers. In mental mediums, there was the assumption of a fictitious "influence of external agencies," and of a "promptness of the brain to give a reflex action of these impressions back upon the outward world through the medium of the automatic apparatus in the bodily frame, or through the odylic agent . . . ." (Rogers 1853:176). The medium's brain, Rogers believed, "may, under certain circumstances in its action, assume any personality, from that of a divinity to that of a toad . . ." (Rogers 1853:171).

The well-known American clairvoyant Andrew Jackson Davis (1826–1910) (see photo) believed in different agencies. In his book *The Present Age and Inner Life* (1853) Davis wrote that "owing to the extraordinary attributes of man's mind, many experiences are by some individuals regarded as spiritually originated; which in truth, are only caused by the natural laws of our being . . . ." (Davis 1853:160f). Davis believed that 40% of the phenomena were due to discarnate spirits. The remaining possibilities included a variety of medical explanations, with 18% being accounted for by what would later be referred to as the psychic powers of the living. This living agency, Davis wrote, was due to "vital electricity" projected by the medium's body, and

Andrew Jackson Davis

by clairvoyance. Like Davis, other writers accepted both spirit and human agency as explanations for the phenomena of mediumship. One of them, Unitarian minister Adin Ballou (1803–1890), believed the phenomena "sometimes proceed from spirits in the flesh," and sometimes from discarnate spirits (Ballou 1853:78). For Ballou the mind of the medium was open to "mesmeric and psychological influences, from controlling minds near them . . . ." (Ballou 1853:61). However, he believed it was possible to distinguish human from discarnate agency. Another author who accepted the notion of mixed agency was the writer Epes Sargent (1812–1880). In his book *Planchette; or, the Despair of Science* he wrote:

> It is not unlikely that many of the minor phenomena, attributed with sincerity by many partially developed mediums to spirits, may be produced by the unconscious exercise of spiritual powers latent in the individual; while other phenomena are of so extraordinary a character that the more rational explanation may be found in the theory of the application of an external spiritual intelligence or force. (Sargent 1869:233)

In Germany, Justinus Kerner (1786–1862) (see photo), famous for his work with the "Seeress of Prevorst," argued that the spiritistic interpretation of table turning and mediumistic communication was misleading. In 1853, Kerner published *Die Somnambulen Tische*, a booklet in which he described experiences and experiments that had been performed on table turning and related phenomena. Some of the examples presented by Kerner concerned allegedly successful transmissions of mental suggestions by sitters, including cases of xenoglossy. Kerner argued that the peculiar movements of tables and

Justinus Kerner

other objects were caused by the "human nerve spirit" of the sitters (Kerner 1853:52), a fluidic force expressed by their organisms. In addition, he attributed the received communications to a subconscious psychic aspect of this nerve spirit that would enable thought-transference and clairvoyance.

Similar thoughts were expressed by noted physician Carl Gustav Carus (1789–1869) (see photo). Along with Gotthilf Heinrich von Schubert (1780–1860), Carus is often regarded as one of the first writers who had laid foundations for exploring

Carl Gustav Carus

and understanding the subconscious layers of the human psyche. In his book *Über Lebensmagnetismus* (1857), Carus stated that table turning, rappings, and mediumistic communications "can be reduced to the rise of the subconscious into the region of the conscious" (Carus 1857:222). Like Kerner, he attributed table movements to influences of animal magnetism. The mediums would be especially prone to utilize this "stream of innervation" (Carus 1857:225) subconsciously. If veridical elements were contained in the received communications, Carus attributed them to the divining faculties of the human subconscious, sometimes also displayed in dreaming. He regarded the spreading of mediumship and sitter groups as an indication of an epidemic of regressive human madness.

The German zoologist Maximilian Perty (1804–1884) was another critic of belief in spirit communication in mediumship. Like Bruno Schindler (1797–1859) (Schindler 1857) before him, in *Die Mystischen Erscheinungen der Menschlichen Natur* (1861/1872), Perty provided an alternative account of certain supernormal phenomena to that of such previous authors as Joseph Görres (1776–1848) (Görres 1836–1842), who held that they were usually caused by the supernatural agency of angels and demons, while Schindler and Perty proposed they were usually due to the vital or psychic forces of living persons.

Anticipating Carl du Prel's (1839–1899) philosophy of dissociation in somnambulism (du Prel 1885), Perty argued that the "guides" of somnambulistic mediums, often assuming the appearance of deceased loved ones, were usually dramatized personifications from the somnambulist's own psyche (Perty 1861/1872(1):244ff). Rather than suggesting evidence of spirit identity, Perty held that supernormal knowledge emerging in "spirit guides" was due to the somnambulist's own unconscious clairvoyance or "magic excitation." In a discussion of the literary products of Andrew Davis, the famous "Poughkeepsie Seer" and precursor of modern spiritualism we have mentioned above, Perty found them to be a mixture of inspiration and obvious errors, and he concluded that they were a "remarkable product of somnambulist ecstasy" (Perty 1861/1872(1):339), combining certain cultural influences that Perty presumed Jackson must have been exposed to, but with glimpses of genius and supernormal knowledge. Similarly, in a contribution to the problem of demonic possession, Perty held that, in most cases, the possessing entities were "products of the own divided [entzweiten] psyche of those concerned" (Perty 1861/1872(1):341), even if apparent possession episodes were accompanied by supernormal physical or mental phenomena. The second volume of *Die Mystischen Erscheinungen* opens with a chapter dedicated to the phenomena of spiritualism, i.e. table rapping, and automatic writing and speaking (Perty 1861/1872(2):1–77). Perty criticized spiritualists claiming that a "spirit's" opposition to views held by a medium counted as evidence for the alleged spirit's autonomy, for our normal dream life was replete with instances of opposing viewpoints of the dream self and the waking self (Perty 1861/1872(2):10–11). Perty therefore concluded: "In a considerable number of cases, the so-called spirits turn out to be products of the magically excited psyche of the mediums, appearing to be autonomous extraneous beings" (Perty 1861/1872(2):60).[5]

Concluding Remarks

The material discussed in this note could be extended. It shows that explanations of veridical elements arising from the living medium rather than from discarnate influence preceded the founding of the SPR. Moreover, such ideas had been put forward during the early days of the SPR without being influenced by the SPR (e.g., von Hartmann 1885a, 1885b). Nonetheless, the early ideas were often not exactly equivalent to those held by SPR workers and later writers engaged in controversies (e.g., Gasperini 2011) and discussions of the so-called hypothetical construct referred to as super ESP (e.g., Gauld 1982, Hart 1959, Sudduth 2009). Instead early ideas were frequently associated with unorthodox concepts of force not discussed by the SPR workers who wrote about mental mediumship.[6] In addition, although earlier writers knew about such processes as those akin to unconscious cerebration, the SPR (and later) workers laid more emphasis on less physiological conceptions of subconscious processes that had a wider scope. The later interest took place in the context of observations of and research on dissociation and the subconscious mind during the late 19th century (Alvarado 2002, Crabtree 1993, Ellenberger 1970). Similar to secondary personalities and the states observed in hypnosis, in hysterical patients, and in other circumstances, mediumship was discussed by some as an example of dissociation and of the existence of the subconscious mind.

Although other authors have discussed some of these ideas of the living medium as the source of veridical information before the foundation of the SPR (see the reviews of Aksákow 1894, Crabtree 1993, Podmore 1902), or early after its foundation and independently from it (von Hartmann 1885a, 1885b), the literature we present is largely unknown among current writers. We hope that this note serves as a reminder that the concept of living agency in the interpretation of the phenomena of mediumship has a history that extends back to the middle of the 19th century.

Notes

[1] On spiritualistic mesmerism, see Crabtree (1993:196–212). See also Alvarado (2009).

[2] Discussions of concepts of force in relation to mediumship include the works of Alvarado (2006, 2008b), Crabtree (1993), and Podmore (1902). Later authors who discuss such forces in the 19th century include Collyer (1871), Cox (1879), De Rochas (1897), and von Hartmann (1885a, 1885b). For example, referring to the medium's "nervo-vital fluid" affecting tables and other objects it was stated: "When this fact is admitted, there is no necessity of ascribing the phenomena to spiritual causes" (Collyer 1871:110). It should be kept in mind that believers in discarnate agency also accepted the existence of forces to explain the phenomena of mediums (e.g., Ballou 1853, Kardec 1863). But they postulated that spirits made use of such forces to produce both mental and physical phenomena.

[3] This review was later included in his influential work *Animismus und Spiritismus* (Aksákow 1894).

[4] According to Baron Karl von Reichenbach (1849/1851), Od was a universal force similar to animal magnetism, that could be found in minerals, crystals, and living organisms including the human body, but that was also generated by several other means such as heat, light, friction, electricity, chemical reactions, and magnetism. The concept was used by many in the mid-19th century to explain mediumship and other psychic phenomena (e.g., Brittan & Richmond 1853, Guppy 1863, Mayo 1852, Rogers 1853, see also Alvarado 2008a). Although in general spiritualists did not like phenomena explained in terms of Od—the writings of Samuel B. Brittan (c. 1815–1883) is one example (Brittan & Richmond 1853)—there was widespread interest in the topic, as is evident in articles published in 19th-century spiritualist periodicals (e.g., Gregory 1872, Howitt 1860). However, von Reichenbach himself didn't relate the concept of Od to mediumistic communications. When he wrote about mediumistic processes, he was more interested in physical phenomena (see, e.g., his presentation of the experiments he had performed on table turning in von Reichenbach 1867:108–148). On Reichenbach and Od, see Nahm (2012).

[5] Perty (1861/1872(2):65) also provided a brief discussion of the role of cryptomnesia in mediumship. That is, mediums seem to generate new information whereas they only reproduce forgotten memories without being aware of doing so.

[6] The pre-SPR British literature on mediumistic forces in general (e.g., Cox

1879, Crookes 1874, Guppy 1863), and of ideas of forces involved in mental mediumship in particular (Alvarado 2008c) notwithstanding, the discussions of mental mediumship published by SPR leaders followed a mentalistic approach. Where force was discussed it was in the context of physical phenomena, Lodge's (1894:334–335) and Myers' (1903(2):544–546) works being examples.

Acknowledgment

The authors wish to thank Nancy L. Zingrone for useful editorial comments.

References

Aksákow, A. (1886). Kritische Bemerkungen über Dr. Eduard von Hartmann's Werk: "Der Spiritismus." Historische Ueberschau der anti-spiritistischen Theorien. *Psychische Studien, 13,* 62–76, 109–126.

Aksákow, A. (1894). *Animismus und Spiritismus* (second edition). Leipzig: Mutze.

Almeder, R. F. (1992). *Death and Personal Survival: The Evidence for Life after Death.* Lanham, MD: Littlefield Adams.

Alvarado, C. S. (2002). Dissociation in Britain during the late nineteenth century: The Society for Psychical Research, 1882–1900. *Journal of Trauma and Dissociation, 3,* 9–33.

Alvarado, C. S. (2003). The concept of survival of bodily death and the development of parapsychology. *Journal of the Society for Psychical Research, 67,* 65–95.

Alvarado, C. S. (2006). Human radiations: Concepts of force in mesmerism, spiritualism, and psychical research. *Journal of the Society for Psychical Research, 70,* 138–162.

Alvarado, C. S. (2008a). Further notes on historical ideas of human radiations: II. Od and psychometry. *Psypioneer, 4,* 63–69. http://www.woodlandway.org/PDF/PP4.3March08.pdf

Alvarado, C. S. (2008b). Further notes on historical ideas of human radiations: III. The perispirit and mediumistic forces. *Psypioneer, 4,* 85–92. http://www.woodlandway.org/PDF/PP4.4April08.pdf

Alvarado, C. S. (2008c). Further notes on historical ideas of human radiations: IV. Mental phenomena. *Psypioneer, 4,* 119–126. http://www.woodlandway.org/PDF/PP4.5May2008.pdf

Alvarado, C. S. (2009). Mesmeric mediumship: The case of Emma. *Psypioneer, 5,* 43–51. http://www.woodlandway.org/PDF/PP5.2February09.pdf

Ballou, A. (1853). *An Exposition of Views Respecting the Principal Facts, Causes, and Peculiarities Involved in Spirit Manifestations* (second edition). Boston: Bela Marsh.

Barkas, T. (1862). *Outlines of Ten Years' Investigations into the Phenomena of Modern Spiritualism, Embracing Letters, Lectures, & Etc.* London: Frederick Pittman.

Braude, S. E. (2003). *Immortal Remains.* Lanham, MD: Rowman & Littlefield.

Brittan, S. B., & Richmond, B. W. (1853). *A Discussion of the Facts and Philosophy of Ancient and Modern Spiritualism.* New York: Partridge & Brittan.

Carus, C. G. (1857). *Über Lebensmagnetismus.* Leipzig: Brockhaus.

Chevillard, A. (1869). *Études Expérimentales sur le Fluide Nerveux et Solution Definitive du Problème Spirite.* Paris: Victor Masson.

Collyer, R. H. (1871). *Mysteries of the Vital in Connexion with Dreams, Somnambulism, Trance, Vital Photography, Faith and Will, Anaesthesia, Nervous Congestion and Creative Function* (second edition). London: Henry Renshaw.

Cox, E. W. (1879). *The Mechanism of Man: An Answer to the Question What Am I: A Popular Introduction to Mental Physiology and Psychology: Volume 2: The Mechanism in Action.* London: Longman.

Crabtree, A. (1993). *From Mesmer to Freud: Magnetic Sleep and the Roots of Psychological Healing.* New Haven, CT: Yale University Press.

Crookes, W. (1874). *Researches in the Phenomena of Spiritualism.* London: J. Burns.

Davis, A. J. (1853). *The Present Age and Inner Life: A Sequel to Spiritual Intercourse: Modern Mysteries Classified and Explained.* New York: Partridge and Brittan.

De Gasparin, A., Count (1854). *Des Tables Tournantes: Du Surnaturel en Général et des Esprits* (two volumes). Paris: E. Dentu.

De Rochas, A. (1897). Les expériences de Choisy-Yvrec (prés Bordeaux) du 2 au 14 octobre 1896. *Annales des Sciences Psychiques, 7,* 6–28.

Dods, J. B. (1854). *Spirit Manifestations Examined and Explained: Judge Edmonds Refuted; or, an Exposition of the Involuntary Powers and Instincts of the Human Mind.* New York: De Witt & Davenport.

du Prel, C. (1885). *Die Philosophie der Mystik.* Leipzig: Ernst Günther.

Ellenberger, H. F. (1970). *The Discovery of the Unconscious: The History and Evolution of Dynamic Psychiatry.* New York: Basic Books.

Fontana, D. (2005). *Is There an Afterlife?* Hants, UK: O Books.

Gasperini, L. (2011). Criptestesia o ipotesi spiritica? Charles Richet ed E. Bozzano a confronto. *Luce e Ombra, 111,* 113–126.

Gauld, A. (1982). *Mediumship and Survival.* London: Heinemann.

Görres, J. von (1836–1842). *Die christliche Mystik* (four volumes). Regensburg & Landshut: G. Joseph Manz.

Gregory, W. (1872). Reichenbach's magnetic flames. *The Spiritualist,* November 15, 5–6.

[Guppy, S.] (1863). *Mary Jane; or, Spiritualism Chemically Explained.* London: John King. [Samuel Guppy wrote with the pseudonum "A Child at School".]

Hart, H. (1959). *The Enigma of Survival: The Case for and against an After Life.* Springfield, IL: Charles C. Thomas.

Hartmann, E. von (1885a). *Der Spiritismus.* Leipzig/Berlin: Friedrich.

Hartmann, E. von (1885b). Spiritism. *Light, 5,* 405–409, 417–421, 429–432, 441–443, 453–456, 466–470, 479–482, 491–494.

Hodgson, R. (1892). A record of observations of certain phenomena of trance. *Proceedings of the Society for Psychical Research, 8,* 1–167.

Howitt, W. (1860). Correspondence: Reichenbach on Od. *Spiritual Magazine, 1,* 566–567.

Kardec, A. (1863). *Spiritisme Expérimental: Le Livre des Médiums ou Guide des Médiums et de Évocateurs* (sixth edition). Paris: Didier.

Kerner, J. (1853). *Die Somnambulen Tische.* Pfullingen: Johannes Baum.

Lafontaine, C. (1866). Avis aux spirites, la cause des effets dits spirites est dans le système nerveux. *Magnétiseur: Journal de Magnétisme Animal, 7,* 113–116.

Leaf, W. (1890). A record of observations of certain phenomena of trance (3). Part II. *Proceedings of the Society for Psychical Research, 6,* 558–646.

Lodge, O. J. (1894). Experience of unusual physical phenomena occurring in the presence of an entranced person (Eusapia Paladino). *Journal of the Society for Psychical Research, 6,* 306–336.

Mahan, A. (1855). *Modern Mysteries Explained and Exposed in Four Parts.* Boston: J. P. Jewett.

Mayo, H. (1852). *Popular Superstitions, and the Theories Contained Therein, with an Account of Mesmerism* (from the third London edition). Philadelphia: Lindsay and Blakiston.

Morin, A. S. (1858). Review of *Le Monde Spirituel ou Science Chrétienne de Communiquer Intimement avec les Puissances Célesres et les Ames Heureuses* by G. de Candemberg. *Journal du Magnétisme, 17*(2), 90–103.

Morin, A. S. (1860). *Du Magnétisme et des Sciences Occultes.* Paris: Germer Baillière.

Myers, F. W. H. (1903). *Human Personality and Its Survival of Bodily Death* (two volumes). London: Longmans, Green.

Nahm, M. (2012). The sorcerer of Cobenzl and his legacy: The life of Karl Ludwig Baron von Reichenbach, his work and its aftermath. *Journal of Scientific Exploration, 26*(2), 381–407.

Oldfield, T. (1852). *"To Daimonion," or, The Spiritual Medium.* Boston: Gould and Lincoln.

Perty, M. (1872). *Die Mystischen Erscheinungen der Menschlichen Natur* (two volumes) (second enlarged edition). Leipzig/Heidelberg: Winter. [Original work published 1861]

Podmore, F. (1902). *Modern Spiritualism: A History and a Criticism* (two volumes). London: Methuen.

Radclyffe-Hall, M., & Troubridge, U. (1919). On a series of sittings with Mrs. Osborne Leonard. *Proceedings of the Society for Psychical Research, 30,* 339–554.

Reichenbach, C. von (1851). *Physico–Physiological Researches on the Dynamics of Magnetism, Electricity, Heat, Light, Crystallization, and Chemism, in Their Relation to Vital Force.* New York: J. S. Redfield. [First published in German 1849]

Reichenbach, K. von (1867). *Die Odische Lohe und Einige Bewegungserscheinungen als Neuentdeckte Formen des Odischen Princips in der Natur.* Vienna: Braumüller.

Rogers, E. C. (1853). *Philosophy of Mysterious Agents.* Boston: J. P. Jewett.

Samson, G. W. (1860). *Spiritualism Tested; or, the Facts of Its History Classified.* Boston: Gould and Lincoln.

Sandby, G. (1850). Review of M. Alphonse Cahagnet's *Arcanes de la Vie Future Dévoilés, &c. Zoist, 7,* 414–431.

Sargent, E. (1869). *Planchette; or, the Despair of Science.* Boston: Roberts Brothers.

Schindler, H. B. (1857). *Das Magische Geistesleben. Ein Beitrag zur Psychologie.* Breslau: Korn.

Sudduth, M. (2009). Super-psi and the survivalist interpretation of mediumship. *Journal of Scientific Exploration, 23*(2), 167–193.

Journal of Scientific Exploration, Vol. 26, No. 4, pp. 867–880 2012   0892-3310/12

ESSAY

Seeking Immortality?
Challenging the Drug-Based Medical Paradigm

HENRY H. BAUER

2012 Tim Dinsdale Award Address
Society for Scientific Exploration, Boulder CO, June 2012

Introduction to Henry Bauer—I am very honored this morning to introduce the recipient of the 2012 Tim Dinsdale Award, Professor Henry H. Bauer. I trust that everyone in this room is aware of the enormous contributions that Henry has made to Anomalistics and to science in general. He was not only one of the Founders of the Society for Scientific Exploration, but also one of its greatest contributors. We owe him very much.

Let me say a few words about his background. Henry was born in Vienna in 1931, but went through school and university in Sydney, Australia. After a Bachelor of Science (given with Honors) he proceeded to receive a Master of Science and Doctorate from the University of Sydney, which then kept him on for seven years to teach in the Department of Agricultural Chemistry. After that he taught at the University of Kentucky for another twelve years, and then went to Virginia Tech, where he spent 21 years before retirement in 1999, and where he is still Professor Emeritus. He was Dean of Arts and Sciences at Virginia Tech for eight years, an experience that led him to write (under a pseudonym) a book called *To Rise Above Principle*. In addition to his regular appointments, he had visiting appointments at the University of Michigan, the University of Southampton, and Rikagaku Kenkyusho in Tokyo.

I will not dwell on his many contributions to electrochemistry, but will simply note that he was author or co-author of three books and many articles and presentations on that subject. Already by the 1970s, however, Henry had begun to turn his attention to science studies, and in particular to the role that anomalies and unconventional science played in scientific thought. This led to research and eventually to books on Immanuel Velikovsky and the Loch Ness "Monsters" and most recently to a strongly dissenting book on the relationship of HIV and AIDS. In addition he wrote several books on the social relations of science, including one with the intriguing title *Scientific Literacy and the Myth of the Scientific Method*. Although a complete gentleman in all his activities, Henry never shrank from confrontation with and opposition to what he considered bad science. I think the phrase "uncompromising integrity" is appropriate here.

I need to note that Henry has written yet another book, which will be available shortly: *Dogmatism in Science and Medicine: How Dominant Theories Monopolize Research and Stifle the Search for Truth.*

And finally, in awarding Henry this Dinsdale Prize, it is impossible not to underline the singular appropriateness of this prize, since he is not only, like Tim Dinsdale, the author of a book on Loch Ness, but in 1978 was made an Honored Companion of the Loch Ness Explorers. And finally I might add that it gives me special pleasure to give this award to a very good friend.

I give you Henry Bauer. Professor Bauer, would you stand and receive the citation for this award?   —Ron Westrum

I appreciate and value this honor more than I can find words to say. This Society has been my intellectual home, and to be recognized in this way by this group is meaningful beyond words.

By serendipity or synchronicity, what I've been looking into recently happens to fit perfectly into the theme of this year's conference, challenging mainstream paradigms. My challenge is to the routine treatment of chronic ailments with drugs, which focuses on symptoms rather than causes and reflects an implicit, unwitting presumption that lifespan can be extended indefinitely.

Some of the following assertions may seem excessively iconoclastic (I'll be rather disappointed if they don't), but everything is based solidly on mainstream sources that have not been seriously questioned. I will give here just a few citations on critical points, but all the factual claims can be verified in a rather lengthy bibliography available at my website (Bauer 2012).

When we don't feel well, it's broadly speaking for one of two quite distinct reasons: Either we have contracted an infection, in other words we've been attacked from outside by a bacterium or a virus or a parasite; or we are experiencing a non-infectious, internally generated problem. Some of those are temporary, and my concern here is with those that are chronic rather than episodic—cardiovascular disease, for example. A chronic ailment bespeaks some disturbance of the intricately interlocking, system-regulating processes of signals, reactions, and feedbacks that normally keep us functioning properly—the phenomenon of homeostasis, physiology keeping itself stable.

Since illnesses arise in these two quite different ways, it seems on first principles that they need to be addressed in quite different ways. Instead, present-day practices apply the same approach in both cases, namely, treatment with drugs.

That's a plausible response to attacks from outside by infectious diseases, and indeed those have been largely overcome by antibiotics, substances able to kill intruders without causing intolerable collateral damage to ourselves. The success of drug treatment of infections is owing largely to the facts that intruders are a specific identifiable cause of illness; the physiologies of bacteria and of parasites are sufficiently different from ours that chemicals can be found that kill the intruders with considerable selectivity and minimum damage to ourselves; treatment can be of quite short duration, days or a few weeks, so that collateral damage done by the drugs is not so serious that it cannot be repaired. Such repair is often needed, for instance because antibiotics also kill the beneficial bacteria in our digestive tracts which are the immune system's first line of defense.

By contrast, treating non-infectious disease, chronic disease, with drugs means lifelong consumption of medication and the accumulation of damage from "side" effects. Yet this is the standard approach nowadays to conditions described as chronic diseases.

With infectious diseases effectively under control, medicine became increasingly preoccupied with the main non-infectious causes of death—the failure of organs and rogue mutation of cells to generate cancers. These are malfunctions for which no single specific cause has been found. Indeed, as a matter of empirical fact, the unsuccessful decades-long search for causes suggests that there is no single one. Cancers are presently thought to be caused by a series of necessary but individually insufficient steps; even when one of them appears to be particularly effective, it is still not in itself decisive: For example, no one disputes that smoking plays a major role in the incidence of lung cancer—speaking statistically, that is; but smoking is neither a necessary nor a sufficient cause of lung cancer, for some people who have never inhaled tobacco smoke contract lung cancer while some heavy smokers live to the ripest old age. When it comes to ailments of heart and arteries, some 243 "risk factors" have been reported (Institute of Medicine 2010).

It would seem clearly irrational to imagine that conditions with such a range of partial apparent causes could be rectified by administering a few drugs, yet that is the current practice: Cardiovascular disease is treated with a variety of cholesterol-lowering drugs, blood-pressure–lowering drugs, blood thinners, beta-blockers, and more.

Now I was actually wrong to say that these non-infectious conditions, chronic and not episodic conditions, have no single specific cause. They do have a single cause. They are brought on by getting older. When someone dies from heart disease or cancer, it's regarded quite properly as death from natural causes. So present-day medical practice prescribes drugs as though that could stave off or even reverse what is a normal progression. The implicit but unacknowledged ambition is to prevent aging.

This ambition—one might call it hubris or chutzpah—is facilitated and masked by semantics: Conditions of aging are talked of as though they were illnesses; normal changes with age are labeled **ab**normal. The ubiquitous term *cardiovascular disease* implies that this is similar to other diseases including infectious diseases. "Hypertension," "too high," obviously describes an abnormal state. "High cholesterol" again implies something abnormal, unnatural, undesirable, *sick*—and "bad cholesterol" sounds even worse and more unnatural. "Dysfunction," as in erectile, means not working properly. "Pre-diabetes" suggests that you will later become ill; yet the actual fact is just that the level of blood sugar under some circumstances happens to be a bit higher than is average for some population of young, healthy individuals.

So chronic diseases are at least partly created or invented by definition. Further, because there is no single identifiable cause and since the symptoms are those of natural aging, taking drugs to treat or prevent chronic ailments— organ failures, cancers—seems very unlikely to work. But that is just arguing from first principles. Since drugs have actually been used to treat chronic diseases for several decades, there ought to be empirical evidence with which to check this a priori reasoning.

There is not.

Has it been the case that statins or blood thinners or anti-hypertensive drugs have prolonged life, with acceptable quality of that prolonged life?

There are no conclusive data-based answers.

The problem is that the proper question—the effect of treatment on all-cause mortality and quality of life—could only be answered by literally impossible studies. For each condition and each treatment of it, one would need to follow for decades a large number of individuals in two groups, a treated group and an untreated control group, the two groups being matched individual by individual for every variable that might affect the condition and its outcome—age, sex, genes, environment, lifestyle including diet.

Such trials being impossible, evidence has been sought indirectly and circumstantially via *biomarkers* or *surrogate markers*: Measurable characteristics whose magnitudes are taken to reflect the status of the condition being treated. For instance, as measures of cardiovascular disease (or risk of it), commonly used biomarkers include blood pressure, various cholesterol-related numbers, proteins associated with inflammation, and blood-clotting time.

But how do we know that any given surrogate marker actually reflects the state of the so-called disease?

We don't.

To know that a surrogate marker is inextricably linked with whatever actually *causes* a particular ailment would require one of those literally impossible clinical trials. Surrogate markers are chosen because they've been found to be associated in some manner with a particular ailment. Surrogate markers are correlated with the condition they purportedly measure; but correlation never proves causation.

The incidence of organ failures and of cancers increases with age. Mortality increases with age. Blood pressure increases with age. The incidence of strokes and of heart attacks increases with age. As we get older, hearing deteriorates. The incidence of dementia increases with age. The incidence of type-2 diabetes increases with age. Arthritis becomes more common as we age.

Everything that increases with age correlates inevitably with everything else that increases with age. Such correlations do not demonstrate that one factor causes the other. It is worth always bearing in mind the example given by Huff (1954) in *How to Lie with Statistics*: Over many years, there was a perfect correlation between the salaries of Protestant ministers in New England and the price of Cuban rum. Did this mean that ministers were receiving salary raises so that they could afford to consume Cuban rum? Or was the ability of the ministers to pay more leading to greater demand for Cuban rum and therefore higher prices?

Of course not. Everything that has a dollar value attached to it experiences increases over time because of inflation, and so—*mutatis mutandis*—the dollar values of many things correlate with one another. But those correlations do not prove that one thing is the cause of the other.

Blood pressure correlates with organ failure, stroke, heart attack, cancer, arthritis, erectile dysfunction, hearing loss, dementia—among other things. That does not demonstrate that high blood pressure is a perfect marker for any of those things, still less that it actually causes any of those things. Correlation never proves causation. Yet present-day medical practice is to administer pressure-lowering medication when blood pressure exceeds a level that has been designated as "hypertension"—designated arbitrarily, based on opinion that mistakes correlation for causation.

Some months ago an entrepreneurial research group reported a correlation between loss of hearing and dementia. Shortly after that my local paper had a full-page advertisement in which a group of audiologists urged everyone to have their hearing tested and if necessary to get hearing aids in order to stave off Alzheimer's (Beltone 2012):

> If hearing loss goes untreated, a condition called "auditory deprivation" occurs, this has been confirmed by scientific studies. Seniors with hearing loss are significantly more likely to develop dementia over time than those

who retain their hearing, a study by Johns Hopkins and the National Institute on Aging researchers suggests. The findings, researchers say, could
lead to new ways to combat dementia, a condition that affects millions of
people worldwide.

Although the reason for the link between the two conditions is unknown, the investigators suggest that a common pathology may underlie
both or that the strain of decoding sounds over the years may overwhelm
the brains of people with hearing loss, leaving them more vulnerable to
dementia. They also speculate that hearing loss could lead to dementia by
making individuals more socially isolated, a known risk factor for dementia
and other cognitive disorders.

But there's good news! Whatever the cause, the scientists report, their
finding may offer a starting point for interventions—even as simple as hearing
aids—that could delay or prevent dementia by improving patients' hearing.
(www.hopkinsmedicine.org)

That may strike you as absurd. It should. Yet the evidence that high
blood pressure causes cardiovascular disease is of exactly the same ilk as
the evidence that hearing loss causes dementia—or for that matter, that
dementia causes hearing loss. It is also worth noting that these absurd
speculations came from researchers at Johns Hopkins University and
the National Institute on Aging. Incompetence in statistics is pervasive,
including in medical matters. Note too the suggestion that "a common
pathology" may underlie both dementia and hearing loss: Yes, a common
factor does underlie them, but it's not a pathology, it's just the process of
getting older.

Since surrogate markers are correlations and not proven causes, the
question becomes how valid a measure they might be of the chronic diseases
with which they are associated. An Institute of Medicine (2010) report has
the explicit aim of defining conditions for use of surrogate markers in order
to limit the potential damage from using markers that are not strictly valid.

The report finds that even the most widely accepted surrogates are far
from universally valid. Thus tumor size is *not* a valid measure of cancer
progression or guide to prognosis—even as it is commonly used as such
a measure. Blood pressure and cholesterol, commonly used to indicate
risk or progression or prognosis of heart disease are not valid measures of
heart disease. A recent article in the *British Medical Journal* says all this
quite plainly: "There are no valid data on the effectiveness" of "statins,
antihypertensives, and bisphosphanates" (the last are prescribed against
osteoporosis) (Järvinen et al. 2011).

So the contemporary paradigm of drug-based treatment of chronic
ailments is flawed in a variety of interlocking ways. The theoretical basis
is misguided:

> Aging is treated as an illness.
> Illness is measured by surrogate markers that are not valid.

In application of that paradigm there are a number of technical flaws as well:

> Drugs are approved on the basis of surrogate markers that are not valid; and there are a number of other flaws in the approval process.
> It's wrongly presumed that drugs can be specific.
> There are many deficiencies in clinical trials.
> Statistical analyses are often made incorrectly.

The approval of drugs by the Food and Drug Administration (FDA) relies on data from surrogate markers; and the FDA requires only two statistically significant trials of at least 6 months duration as supporting evidence. This procedure was originally introduced as "accelerated approval" during the panicked early days of the AIDS era, when activists were clamoring for treatment. An integral requirement of the new procedure: There was to be monitoring of the actual effect of treatment once a drug came into general use. However, that requirement has not been fulfilled, for about three-quarters of all drugs. That is owing in some part to the fact that the costs of the approval process are borne by the company making the request for approval, and the law specifying such payments precludes the FDA from using those funds for post-approval monitoring.

Altogether, there is no system in place to ensure that adverse "side" effects of drugs are actually reported, let alone promptly; and there exist no specific guidelines about when a drug should be withdrawn from the market. Physicians and hospitals are requested to alert the manufacturers or the FDA of adverse events, but it is not a formal requirement and there are no guidelines or specified criteria to help doctors determine what should be reported. In any given case of a patient's condition worsening, the physician may be uncertain whether this can be clearly ascribed to the medication or should be interpreted as a failure of the medication to defeat the illness, which would seem the natural immediate presumption. Even when side effects are reported to manufacturers, there is no clearly specified obligation on the manufacturer to alert the FDA. The Adverse Events Reporting System of the FDA is estimated to pick up only about 10 percent of actually occurring negative side effects, in other words perhaps 90 percent of all adverse drug reactions go unreported (Bowser 2011).

In addition to the reliance on invalid surrogate markers, there are quite a number of reasons why clinical trials have not delivered reliable knowledge about the efficacy and safety of drugs used in the manner in which they are actually prescribed:

Conflicts of interest may bias the initial trials. Clinical trials are increasingly being done by Contract Research Organizations (CROs) (Mirowski 2011). Their business is more likely to thrive if their clients get the results they want. Predisposition on the part of drug companies and of CROs to expect that a new drug will be effective can influence how trials are carried out and interpreted.

Positive results of the initial clinical trials—on which approval is based—are often not reproducible. Of the 35 most highly cited studies based on biomarkers, only 15 of the tested drugs later achieved even nominal statistical significance in actual use, and only 7 of them had certifiably significant effectiveness (Ioannidis & Panagiotou 2011). That may be owing in part to biased clinical trials, but, further, drug companies are not required to submit the data from *all* trials, only from 2 successful ones; so there may well have been some that failed to deliver significantly positive results. At any rate: Experience shows that *80% of drugs (28 of 35) were approved on the basis of bad data that did not later hold up.*

Trials are not representative of who will be treated in practice.

1. The obvious way to test a drug is with the most clearly ill people. But after a drug has been approved, it will be used with people who are less ill. Drugs to lower blood pressure might be tested with people at Stage 2 or Stage 3 hypertension—above 160 or 180 systolic—and yet after approval those drugs will be prescribed for people whose pressures may be as low as 140 or 130 or even lower.

An actual example: Bisphosphanates achieved a 32% reduction in hip fractures among women aged 65–80 with actual osteoporosis or previous fractures. But if these drugs were administered to all older people at risk of osteoporosis, which corresponds to present-day practices, fewer than 5% of hip fractures would be prevented (Järvinen et al. 2011). Whereas a 32% reduction might make for a good benefit/risk ratio, a 5% reduction would be unlikely to do so.

2. "Off-label" uses of drugs are common. Doctors are permitted to prescribe a drug for any condition at all, not just those for which the FDA approved that drug. About 20% of all prescriptions in the USA were for off-label purposes—about 30% with psychotropic drugs (estimated cost, $26 billion), for instance the antipsychotic Risperdal was used off-label twice as often as on-label. Nearly half of the prescriptions for asthma medicines were off-label. Coumadin (warfarin) is commonly prescribed for hypertension and coronary heart disease even though never approved for such treatment. A review (Radley et al. 2006) found that about three-quarters of off-label applications had little or no scientific support.

Drug companies are forbidden from advertising off-label uses, but they find ways of doing so nevertheless. GlaxoSmithKline agreed to pay $3 billion for illegally marketing Advair, whose annual sales approach $8 billion (Whalen 2012). For illegally promoting off-label uses, Pfizer had been fined $2.3 billion, Eli Lilly $1.4 billion, AstraZeneca $520 million, Bristol-Myers Squibb $515 million, and Novartis $420 million (Pharmalot 2010).

The record suggests that all the leading pharmaceutical companies find it profitable to break the law and pay the fines as just another cost of doing business. Not just by promoting off-label uses: Johnson & Johnson were fined $1.1 billion for hiding risks associated with Risperdal (Health Notebook 2012).

But what has been the human cost of innumerable individuals imbibing inappropriate medications? How much does this illegal quackery contribute to the overall costs of "health care"?

A host of deficiencies have to do with statistics:

Misguided or misleading interpretation: Correlations are persistently taken as demonstrating causes. Risk factors are confused with actual risks. A review in 2011 reported that half of all articles in top-ranked journals—*Science, Nature, Nature Neuroscience, Neuron,* and *Journal of Neuroscience*—used incorrect statistical procedures (Nieuwenhuis et al. 2011). Douglas Altman (1994, 2002) at the Center for Statistics in Medicine in Britain has published about this for more than two decades.

Data mining: Correlations are looked for in existing sets of data— recall the triumphantly published finding of the link between dementia and hearing loss, a perfectly predictable and uninteresting result since both increase with age. If one tries enough combinations, some purely chance associations will appear to be "statistically significant"—under the usual criterion of $p \leq 0.05$, on average at least 1 in every 20 apparent correlations will be spurious (and the other 19 should not be interpreted as reflecting a cause-and-effect relation).

Relative risks are reported instead of absolute risks. If A is a cause of death at a rate of 4 per 100, and if a treatment reduces the rate to 2 per 100, drug companies and the media will typically report a 50% reduction in risk, which seems eminently worthwhile. They might even put it that without the drug, the risk of death is 100% higher. But actually the risk of death is lowered by only 2 per 100, 2%, and that may *not* represent a good bargain with the side effects of any possible treatment.

Even beyond all that, I. J. Good or R. A. J. Matthews (1998, 1999) would point out that the standard frequentist method used in medical statistics (and in social-science research), taking $p \leq 0.05$ as a criterion, can deliver quite misleading results, overestimating the actual significance of any results.

A consequence of accelerated approval and associated flaws is that an increasing number of approved drugs have had to be withdrawn again after less and less time on the market, because of dangerous side effects that had not been noted in the short clinical trials. For example, the anti-diabetic Rezulin, approved in 1997, was withdrawn after only 3 years; the antibiotic Raxar, approved in 1997, withdrawn after only 2 years; the appetite suppressor Redux, approved in 1996, withdrawn after 21 months; the calcium-channel blocker Posicor, approved in 1997 and prescribed against hypertension and angina, was withdrawn after 12 months; the analgesic Duract, approved in 1997, was withdrawn after 11 months; Lotronex, for gastrointestinal disorders, was approved in 2000 but lasted only 10 months before being withdrawn. The Wikipedia list of withdrawn drugs has 2 from the 1950s–1960s, 2 from the 1970s, and 5 from the 1980s. Following the introduction of accelerated approval, 15 had to be withdrawn in the 1990s and 26 in the 2000s.

Current practice is far too cavalier about so-called "side" effects. There is copious evidence that drugs approved for one purpose—reducing blood pressure, say—are likely to have undesired other effects. One illustration of the ubiquity of side effects is the drug industry's continuous efforts to "reposition" drugs, to have them approved for a different purpose than the original one. Purely commercial reasons underlie this practice. Once a drug has been approved, its safety has been established to the satisfaction of the FDA (which, as just pointed out, does not mean the drug is actually safe, however). That hurdle does not have to be overcome again when approval is later sought for some other use of the drug (Ashburn & Thor 2004). Special conferences (Drug Repositioning 2012) reflect the presently high interest in repositioning or "repurposing" drugs.

One example of multiple functions is Propagest, which used to be prescribed for nasal congestion, control of urinary incontinence, priapism, and obesity (Allergy Clinic no date). Since 2005, the Food and Drug Administration has been urging that the drug be withdrawn because of apparent association with increased risk of hemorrhagic stroke, showing the drug's effect on yet another aspect of physiology.

A second example of multiple functions, and still being prescribed, is Cymbalta, originally approved to treat anxiety, depression, diabetic peripheral neuropathy, and fibromyalgia (remarkably enough, all at the same dosage), clear enough evidence in itself that the substance affects a whole range of bodily functions. The drug was later repurposed successfully—under a new name, Duloxetine SUI—as treatment for stress urinary incontinence.

Even more worth pondering than multiple applications is that a given drug may exert opposite side effects in different people at different times: Side effects noted with Cymbalta/Duloxetine include increased urination as well as difficulty urinating; constipation as well as diarrhea; agitation and trouble sleeping but also drowsiness and unresponsiveness (Cymbalta no date).

So the term *side effect* is subtly and dangerously misleading. It implies atypical, not important, as well as unintended. But a drug, a material chemical substance, just does what it does, whether we want it to or not. When the typically short clinical trials turn up only a few unwanted effects in only a few people, that does not mean that other people are not experiencing the same effects in latent, not-yet-observable form. We should rather expect that under sufficiently long treatment, many, most, or even all patients would experience those unwanted "side" effects to an appreciable extent. Drugs prescribed for a chronic condition—blood thinners, pressure-lowering drugs, statins, anti-diabetes drugs—are to be taken forever, so any possible side effects are quite likely to show up sooner or later in quite a lot of people.

Current practice came about because reliance on surrogate markers supplanted feelings of illness on the part of patients and subsequent clinical diagnosis by a physician (Greene 2007). Nowadays we're urged to "know our numbers"—cholesterol, blood sugar, blood pressure, and more. In earlier times, high blood pressure, "hypertension," was diagnosed because of such symptoms as severe headache, dizziness, blurred vision, chest pain. But nowadays hypertension is diagnosed on the basis of blood pressure alone, whether or not there are any symptoms—even though the Institute of Medicine (2010) has acknowledged that blood pressure is not a valid surrogate for cardiovascular disease.

About a century ago, after measuring blood pressure had become routine, it was soon found that it varies over a wide range among manifestly healthy people, and for any given individual it also varies with time of day, temperature, and most notably stress. An average rule of thumb used to be that blood pressure approximates age plus 100 (Graveline n.d.). But even though the normal, natural increase with age is well-known, the official designation of hypertension is independent of age (PubMed Health 2011)!

> Normal blood pressure is . . . lower than 120/80 mmHg most of the time.
> High blood pressure (hypertension) is . . . 140/90 mmHg or above most of
> the time.
> If . . . 120/80 or higher, but below 140/90, it is called pre-hypertension.
> If you have pre-hypertension, you are more likely to develop high
> blood pressure.

Well, of course you are. If your pressure is between 120 and 140, all you have to do is to live sufficiently longer and it will come to exceed 140.

Knowing that blood pressure increases normally with age, it would be rational to define as too high—hypertension—pressures that exceed by (say) 50% what is average *for a given age*. Yet current practice recommends that everyone should have blood pressure lowered to about what is average for people in early middle age; and even lower if you have a condition that is not even known to be caused by high blood pressure (PubMed Health 2011):

> If you have heart or kidney problems, or if you had a stroke, your doctor may want your blood pressure to be even lower than that of people who do not have these conditions.

Under the official definition, then, we all achieve hypertension by about 60 years of age. According to the Institute of Medicine, 75–80% of Americans above age 60 have hypertension. I suggest that it is absurd to regard and treat as sick, about one-third of American adults including 75–80% of seniors, none of whom may have any feeling of being ill. And furthermore, it seems dangerous to administer drugs to be taken lifelong that are intended to counteract this normal age-related increase in pressure. In fact, half a century ago when diuretics were first being marketed to reduce blood pressure, many cardiologists disapproved, calling it a dangerous experiment and pointing out that increasing pressure with age might well be a compensation for the decreased flexibility of arteries, so that more pressure is needed to ensure that enough blood reaches the extremities and all organs (Greene 2007:53 ff.).

To conclude: The present-day drug-centered medical paradigm misinterprets natural accompaniments of aging as illnesses and prescribes lifelong medication with the real risk of debilitating so-called "side" effects. By contrast, an insightful geriatrician has pointed out that the challenges of aging are spiritual rather than medical, and that over-testing and over-treating, merely inconvenient in middle age, border on assault when perpetrated on older people, in whom almost every test is likely to reveal something "out of the ordinary" (Goodwin 1999). Furthermore, since clinical trials rarely if ever enroll people appreciably older than middle age, even the theoretical actual efficacy of drugs taken for chronic conditions is not really known. Enough is known, though, to indicate that older individuals should be cautious about taking even some quite commonly used medications (Beers no date).

Not only chronic ailments are being mis-treated with drugs. Emotional or mental conditions are being treated by drugs that have not been shown to do what they claim to do. Instead, there are nowadays far more Americans chronically disabled and dependent on drugs than were ever in asylums before treatment by drugs superseded other treatment of psychiatric conditions (Whitaker 2010).

It is impossible to estimate how much harm is being done by the paradigm of drug-based treatment of chronic ailments associated with aging. More than a century ago, the Food and Drug Administration was established as a counterweight to the prevalent peddling of panaceas and elixirs of life by confidence men, "snake-oil salesmen." The current practice of promiscuous prescribing of drugs makes it seem as though medicine has regressed by a century or so to those days of snake-oil salesmen—except that nowadays the peddlers are pharmaceutical corporations and not individuals.

But then we have learned recently from the Supreme Court that corporations are really people, so perhaps nothing at all has really changed.

References

Allergy Clinic (no date). http://www.ohioallergy.com/allergy-topics/P/PROPAGEST.html

Altman, D. G. (1994). The scandal of poor medical research. *British Medical Journal*, *308*, 283.

Altman, D. G. (2002). Poor-quality medical research: What can journals do? *JAMA*, *287*, 2765–2767.

Ashburn, T. T., & Thor, K. B. (2004). Drug repositioning: Identifying and developing new uses for existing drugs. *Nature Reviews: Drug Discovery*, *3*, 673–683.

Bauer, H. H. (2012). What's Wrong with Present-Day Medicine. http://henryhbauer.homestead.com/What_sWrongWithMedicine.pdf

Beers Criteria (no date). https://www.dcri.org/trial-participation/the-beers-list

Beltone Hearing Health News (2012). Hearing Loss and Dementia Linked. *Roanoke Times*, 6 February, p. 5 (and on several other occasions). See also press release from Johns Hopkins Medicine citing work by study leader Frank Lin, M.D., Ph.D., assistant professor in the Division of Otology at Johns Hopkins University School of Medicine: http://www.hopkinsmedicine.org/news/media/releases/hearing_loss_and_dementia_linked_in_study

Bowser, B. A. (2011). Certain antibiotics spur widening reports of severe side effects. *PBS Report*, 16 June. http://www.pbs.org/newshour/bb/health/jan-june11/antibiotics_06-16.html, accessed 13 July 2012

Cymbalta (no date). http://health.howstuffworks.com/medicine/medication/cymbalta3.htm http://www.ehealthme.com/ds/cymbalta/agitation http://www.caring.com/questions/does-cymbalta-cause-agitation http://www.drugs.com/sfx/cymbalta-side-effects.html

Drug Repositioning (2012). Indications Discovery and Repositioning Summit—Finding New Tricks for Old Drugs; 13–14 March. http://www.healthtech.com/Drug-Repositioning Second Annual Drug Repositioning Conference; 23–24 October. http://www.drugrepositioningconference.com/index

Goodwin, J. S. (1999). Geriatrics and the limits of modern medicine. *New England Journal of Medicine*, *340*, 1283–1285.

Graveline, D. (no date). Blood pressure and heart disease. http://www.spacedoc.com/blood_
      pressure_heart_disease.htm
Greene, J. A. (2007). *Prescribing by Numbers: Drugs and the Definition of Disease*. Baltimore, MD:
      Johns Hopkins University Press.
Health Notebook (2012). Fraud. *Roanoke Times*, 12 April, p. 4.
Huff, D. (1954). *How to Lie with Statistics*. New York: W. W. Norton.
Institute of Medicine (2010). *Evaluation of Biomarkers and Surrogate Endpoints in Chronic Disease*.
      Washington, DC: National Academies Press.
Ioannidis, J. P. A., & Panagiotou, O. A. (2011). Comparison of effect sizes associated with biomarkers
      reported in highly cited individual articles and in subsequent meta-analyses. *JAMA, 305*,
      2200–2010.
Järvinen, T. L. N., Sievänen, H., Kannus, P., Jokihaara, J., & Khan, K. M. (2011). The true cost of
      pharmacological disease prevention. *British Medical Journal, 342*. doi: 10.1136/bmj.
      d2175
Matthews, R. A. J. (1998). Facts versus factions: The use and abuse of subjectivity in scientific
      research. *European Science and Environment Forum Working Paper*. Reprinted in J. Morris,
      Editor, *Rethinking Risk and the Precautionary Principle*, Butterworth, 2000, pp. 247–282.
Matthews, R. A. J. (1999). Significance levels for the assessment of anomalous phenomena.
      Journal of Scientific Exploration, 13, 1–7.
Mirowski, P. (2011). *Science-Mart: Privatizing American Science*. Cambridge, MA: Harvard
      University Press.
Nieuwenhuis, S., Forstmann, B. U., & Wagenmakers, E.–J. (2011). Erroneous analyses of inter-
      actions in neuroscience: A problem of significance. *Nature Neuroscience, 14*, 1105–1107.
Pharmalot (2010). Here we go again: Novartis fined for off-label marketing, 3 October. http://
      seekingalpha.com/article/228115-here-we-go-again-novartis-fined-for-off-label-
      marketing
PubMed Health (2011). Hypertension. http://www.ncbi.nlm.nih.gov/pubmedhealth/PMH0001502
Radley, D. C., Finkelstein, S. N., & Stafford, R. S. (2006). Off-label prescribing among office-based
      physicians. *Archives of Internal Medicine, 166*, 1021–1026.
Whalen, J. (2012). Glaxo's Asthma Problem. *Wall Street Journal*, 6 July, B1, B2.
Whitaker, R. (2010). *Anatomy of an Epidemic: Magic Bullets, Psychiatric Drugs, and the Astonishing
      Rise of Mental Illness in America*. New York: Crown.

Journal of Scientific Exploration, Vol. 26, No. 4, pp. 881–884, 2012   0892-3310/12

Identity of Shakespeare

I recently came across this *JSE* book review comment on David Roper's website (Roper 2012):

> Roper has provided in this book primary evidence relevant to an enigma of over 400 years standing: The actual identity of "William Shakespeare." (Desper 2009:375)

While it is correct that his book provides this "primary relevance," I actually made this discovery in 2008 (Ferris 2008), where it was and has been in the public forum for a number of years. As you probably know, codes and ciphers are highly suspect in the Shakespeare Authorship debate—even by some (if not most) of those who acknowledge themselves as "Oxfordians." Codes and ciphers, however, have been making headway in scholastic and in popular forums for a while now; progress is somewhat slow, but is picking up some momentum—thanks to David Roper's wonderful contributions.

I thought you might enjoy (or rather, I hope you will enjoy!) my personal graphic (Figure 1) of just one of the finds I have made in Sonnet 76. Using a skip/shift of one transposition equidistant letter sequence (ELS) technique, I placed Sonnet 76 into several arrays, but the particular array I am drawing your attention to is Array 14 (Figure 1). Although the raw probabilities noted below are rough or approximate calculations of deliberate placement with the plaintext of the sonnet, numbers exceeding one million deserve critical attention. As you can see, the word "DEVERE" attached/connected/touching the words: "MY NAME'S" is strong support for probable placement within the plaintext (i.e. put there by intelligent design, or, encrypted, if you will) as well as implying the true writer of Sonnet 76 was Edward de Vere, the Seventeenth Earl of Oxford.

The validity of the letter-string and the sentence it produces ("My name's de Vere") is apparent: It is absolutely there, it can be seen, there is no "torturing" of plaintext accompanied by elaborate interpretations involved. There is no bias controlling the results. The most conservative of all doubters that anyone except the Stratford Shakespeare is responsible for the Shakespeare canon, or the most skeptical of Oxfordians, can get the same results using the simple ELS method. A child can produce the same results every time (pun not intended). In fact, I believe it would be, and is, difficult for others, especially for those with dedicated Stratfordian points of view to dismiss the array graphic with the tired and predictable response: "IT just happened by chance." But how to explain its presence otherwise—wouldn't a mathematical approach be needed to refute what is highly probable, a valid coding of de Vere's name? And not just a mathematical approach, but common sense and/or a reasonable person would be attracted to the array on its own merits.

Again, no "torturing" of the plaintext is necessary; nothing along the lines of trying to explain and/or to figure what is embedded—as is the case with Baconian ciphers.

The "u"s **intended** to be "v"s in actual pronunciation (in Elizabethan times as well as now) have been changed to "v"s, for reasons of clarity.

Sonnet 76

Why is my verse so barren of new pride?
So far from variation or quicke change?
Why with the time do I not glance aside
To new found methods, and to compounds strange?
Why write I still all one, ever the same,
And keepe invention in a noted weed,
That every word doth almost tel my name,
Shewing their birth, and where they did proceed?
O know sweet love I alwaies write of you,
And you and love are still my argument:
So all my best is dressing old words new,
Spending againe what is already spent:
For as the Sun is daily new and old,
So is my love still telling what is told.

"My name's DEVERE" ("My name is DEVERE")
"That every word doth almost tel my name"

Raw Probability Calculations. Total Letters: 448. Letter-String: "DEVERE":
(D = 27) (E = 57) (V = 8) (E = 56) (R = 24) (E = 55) =
[both individual numerator and denominator values are divided by 100 for ease of calculation]
(.27/4.48) (.57/4.47) (.08/4.46) (.56/4.45) (.24/4.44) (.55/4.43) =
.00091010304/7817.4858380774 = 1/8,589,671.1630338 =
.0000001164188 = 1,164,188/10,000,000,000,000,000 =
1 in 8,589,678 = 99.99998835812% (8.5 million to one) raw probability of occurring by chance.

Sonnet 76
Array 14

M Y N A M E I S D E V E R E

WHYISMYVERSESO
BARRENOFNEWPRI
DESOFARFROMVAR
IATIONORQUICKE
CHANGEWHYWITHT
HETIMEDOINOTGL
ANCEASIDETONEW
FOUNDMETHODSAN
OTOCOMPOUNDSST
RANGEWHYWRITEI
STILLALLONEEVE
RTHESAMEANDKEE
PEINVENTIONINA
NOTEDWEEDTHATE
VERYWORDDOTHAL

MOSTTEL **M Y N A M E'S**
HEWINGTH **E** IRBIR
THANDWHE **R** ETHEY
DIDPROCE **E** DOKNO
WSWEETLO **V** EIALW
AIESWRIT **E** OFYOU
ANDYOUAN **D** LOVEA
RESTILLMYARGUM
ENTSOALLMYBEST
ISDRESSINGOLDW
ORDSNEWSPENDIN
GAGAINEWHATISA
LREADYSPENTFOR
ASTHESUNISDAIL
YNEWANDOLDSOIS
MYLOVESTILLTEL
LINGWHATISTOLD

E D W A R D **O** X E N F O R D

Figure 1. Array 14 (skip of 14) of Sonnet 76.

As for me, I am the first to find the encryption in Sonnet 76 (Ferris 2008), which reads: "My name's DEVERE". A more lengthy treatment of this find has been done by David L. Roper in his book *Proving Shakespeare* (Roper 2008) where he states this find is conclusive proof Edward de Vere is the author of the Shakespeare sonnets. Roper's website also contains information on this letter-string, as well as calculated probabilities (Roper 2012), and can be found as Proof Four at http://www.dlropershakespearians. com/ where he states (referring to my find):

> It should therefore be understood that this "autographed" sonnet proves conclusively that Edward de Vere was the poet who wrote Shakespeare's Sonnets. Once chance has been rejected, there is no other explanation. (Roper 2012)

And later:

> Acknowledgement is due to Dr. James Ferris, who first drew attention to de Vere's name in Sonnet 76. (Roper 2012)

I have many more like examples from the plays as well.

JAMES S. FERRIS
drferris68@yahoo.com

References

Desper, C. R. (2009). Review of *Proving Shakespeare: In Ben Jonson's Own Words* by David L. Roper. *Journal of Scientific Exploration 23*(3), 375–376.

Ferris, J. S. (2008). *The Truth Will Out That Ever I Was Borne. The Horatio Code: The Confession of 1623.* http://www.drjsferris.com/

Roper, D. L. (2008). *Proving Shakespeare: In Ben Jonson's Own Words.* Truro, Cornwall, UK: Orvid Publications. 556 pp.

Roper, D. L. (2012). Truth of Authorship. http://www.dlropershakespearians.com/

Journal of Scientific Exploration, Vol. 26, No. 4, pp. 885–891, 2012   0892-3310/12

The AIDS Conspiracy: Science Fights Back by Nicoli Nattrass. Columbia University Press, 2012. 225 pp. $14.99 (Kindle). $34.50 (hardcover). ISBN 978-0231149129.

The official position, the mainstream consensus, is that HIV causes AIDS and that anti-HIV drugs are beneficial. Both are denied by many people: Some of them are eminently qualified to critique the technicalities, others are persuaded by personal experience or that of friends of being "HIV-positive" but healthy, and others again have analyzed the cases presented pro and con by the believers and the disbelievers. To my knowledge, there exists no disinterested analysis of the opposing cases, and books and book reviews tend to be highly polarized. For the present book, fulsome praise has come from those who share Nattrass's belief that HIV causes AIDS; the opposite comes from those who disagree with her. This reviewer disagrees with Nattrass (Bauer 2007a, 2009a), and the reader is thereby warned to be on the alert for bias in this review even as its author strives to focus on verifiable points.

The book's title reflects accurately that the discussion concerns tactics, strategies, and psychological and sociological and political aspects of the to-and-fro between believers and disbelievers. Regarded as insightful, consequently, are such passages as

> Notably, Paula Treichler locates AIDS conspiracy beliefs within what she terms a broader "epidemic of signification" or parallel cultural process in which people generate, reproduce, and perform meanings in an attempt "to understand—however imperfectly—the complex, puzzling and quite terrifying phenomenon of AIDS." (p. 47)

Everything in this book is predicated on the belief that mainstream HIV/AIDS interpretations are unproblematically right, and everything said about disbelievers is predicated on their being completely wrong. To explain the innumerable demonstrable errors thereby introduced would require a volume at least as long as the book itself, so this review addresses only a few salient issues.

The book's title is subtly misleading in alleging a conspiracy (by the disbelievers) and in asserting that "science"—monolithically?

consciously?—is contesting that conspiracy. But it is never established that there is a conspiracy. For example, there is no persisting suspicion on both sides of the Atlantic that "the pharmaceutical industry invented AIDS as a means of selling toxic drugs" (p. 1). Nor is there a common "conspiratorial move" "implying that scientists and clinicians have either been duped by, or are part of, a broader conspiracy to inflict harm." We disbelievers differ among ourselves over many points of detail and have in common just the conviction that the mainstream consensus is wrong. In a few places (e.g., p. 79), Nattrass mentions strong divisions among the "conspirators," yet throughout the book she persistently equates disbelief with Peter Duesberg and "a closely knit group" (p. 8) beholden to him. The book alleges "an organized network of activists" with "linked websites, conferences, papers, books, documentaries, and public relations exercises" (p. 108); in reality, there is nothing organized about the individual blogs and websites and discussion groups, most of which rarely even refer to one another, and Nattrass's own figure (6.1, p. 109) shows only some overlap among the groups to which a few individuals have belonged.

A frequent strategy (for example, see p. 3) is to acknowledge but play down mainstream failings ("although," "admittedly," and the like) and then to assert ("but," "nevertheless," and the like) that the disbelievers are wrong utterly, in wholesale fashion, by "[r]ejecting medical science"; but all we are doing is questioning one set of interpretations. Semantics serves a similar strategy: Believers are "pro-science advocates" "promoting evidence-based medicine" who have answered all the points raised by disbelievers—even as we disbelievers have not encountered answers to our assertions; for instance, that the epidemiology of HIV test results demonstrates that HIV and AIDS are not correlated and that HIV does not behave like an infectious agent (Bauer 2007a). The data we adduce as to the influence of physiological as well as psychological stress, and our citations from mainstream publications concerning the toxicity of AIDS drugs, are not argued against, they are just dismissed by using scare quotes: "stress," "toxic" (p. 5). Disbelievers' claims "have been countered many times by *the scientific community*" (p. 9; emphasis added), as though Duesberg, Mullis, and many other disbelieving scientists were not part of that community. Such sweeping overgeneralizations are frequent, for instance that it is impossible to engage in productive discussions with AIDS denialists (p. 85)—every single one of us, apparently.

So cocksure is this book that it criticizes as "inappropriate" (p. 7) the initiative by South African President Mbeki to stage a debate between believers and disbelievers. What better way for a policymaker to attempt to navigate on a controversial issue? According to Nattrass, any contemporary scientific mainstream consensus should just be taken as revealed truth—which would mean failing to learn from the long history of modern science that progress often or usually entails the replacement or modification of such a consensus (Barber 1961, Bauer 2003, Kuhn 1962/1970, Stent 1972, Hook 2002). Nattrass and her ilk keep denying that there is legitimate debate over HIV/AIDS, even as eminent scientists disagree with her; for example, Nobelist Kary Mullis whose invention is deployed in all studies of HIV "viral load"; or Luc Montagnier, Nobel Prize recipient for discovering HIV, who insists contrary to the mainstream consensus that initially healthy immune systems can stave off

HIV and who agrees with the disbelievers that HIV was never isolated as part of its claimed discovery and the invention of "HIV" tests.

Some points of fact are reported misleadingly by omitting parts of the story:

- ➢ It is true that the coroner initially reported that Christine Maggiore died of AIDS (p. 5). However, it should have been added but was not, that he was sued and the city paid a settlement to avoid court proceedings.
- ➢ Similarly, Chapter 7 recounts the rejection of a Duesberg article by several "peer" reviewers, but neglects to add that the article was eventually published in a well-established, peer-reviewed journal independent of both pro- and con- HIV/AIDS concerns (Duesberg et al. 2011).
- ➢ It is also factually wrong to say that Gallo was cleared of misconduct (p. 112); the Director of the National Institutes of Health just refused to prosecute the case recommended by her own investigating committee (Crewdson 2002).
- ➢ It is further factually wrong that HIV tests are flawed because antigen, antibody, and viral load tests can yield different results (p. 119): The tests are flawed because they have never been shown to detect active infection, nor have they been approved for that purpose (Weiss & Cowan 2004).

➢ Christine Maggiore is said to have increased the risk of infecting her daughter by breastfeeding her (p. 121), yet copious mainstream data show that exclusive breastfeeding *decreases* the chance that the child becomes "HIV-positive" (Bauer 2007b).

➢ Farber's 2006 article in *Harper's* did *not* catapult Duesberg into the limelight (p. 127), it enraged believers by describing the death of a pregnant woman used as guinea pig for antiretroviral drugs; and whistleblower Jonathan Fishbein receives more coverage than Duesberg.

➢ Nattrass's treatment of clinical trials with orphans as subjects is so tendentious as to be wrong (p. 128 ff.); for the authentic story, see http://www.guineapigkids.com

Chapter 2 is misleading by focusing on conspiracy theories that are not held by the overwhelming proportion of serious disbelievers, among them Duesberg and his "closely knit grip" known as Rethinking AIDS as well as the determinedly separate Perth Group. They do not agree with the 30% of black Americans as well as a host of Africans who accept or lend potential credence to the conspiracy theory that HIV was deliberately created as a weapon against black people (p. 12). Nor do Duesberg et al. agree with the far-out notions of William Cooper (p. 23 ff.) or those of Leonard Horowitz or Louis Farrakhan (p. 25 ff.) or of Edward Hooper (p. 29 ff.) or of Boyd Graves (p. 34 ff.). Nattrass focuses on a handful of extremists who are quite unrepresentative of disbelievers (Bauer 2009b).

Chapter 3 is really just about South Africa, where Nattrass resides. She evinces particular animus against former President Mbeki. Readers unfamiliar with HIV/AIDS matters may be puzzled to read that "the symptoms of AIDS . . . [are] diarrhea, tuberculosis, and wasting" (p. 49): Those are the symptoms of *African* AIDS, whereas the original symptoms in the USA and Europe were two fungal infections (thrush or yeast and fungal pneumonia) and Kaposi's sarcoma (purple blotches on skin and other tissues). Disbelievers point out that African AIDS is a quite different phenomenon from the AIDS described in the early 1980s when it first appeared.

Chapter 4 is a paean to David Gilbert, a prisoner in the USA who "cofounded a peer AIDS education initiative" (p. 63). The book also bears the dedication, "For David Gilbert." This is one illustration of Nattrass's praise of anyone who agrees with her views, no matter their lack of pertinent scientific credentials, at the same time as she strives to undercut disbelievers including universally acclaimed retrovirologist Peter Duesberg or Nobelist Kary Mullis because they have not done specifically HIV or

AIDS research (e.g., p. 111). Other totally unqualified people praised for helping "HIV science fight back" (p. 132) are the anonymous Snout and the irresponsible J. T. de Shong (Baker 2010). Nattrass herself is an economist, who simply accepts mainstream HIV/AIDS theory as true. Why she does this is not clear, given that she recognizes, for example, that "gaps in our understanding remain, particularly with regard to precisely how the immune system is destroyed" (p. 79)—which happens to be the central, crucial issue in HIV/AIDS theory. As basis for non-scientists like herself to "exercise some reasonable judgment," she cites "two undeniable and easily grasped facts" (p. 79) that are neither undeniable nor easily grasped—efficacy of antiretrovirals consistent with basic HIV science which, she admits, doesn't understand how HIV kills the immune system! Yet she criticizes Mbeki and AIDS denialism in general for "extreme skepticism toward the science of HIV pathogenesis and treatment" (p. 105). Surely the most extreme skepticism is warranted when that "science" is ignorant about "precisely how the immune system is destroyed" (p. 79).

Chapter 5 continues with the critique of President Mbeki for "questioning HIV science and his conspiratorial move against antiretrovirals." But Mbeki had invited representatives of the mainstream view as well as disbelievers to a panel to advise him. That is hardly a "reject[ion of] scientific expertise" (p. 77). Nattrass admits that the first antiretroviral, AZT, "was plagued by serious side effects" (p. 81). That disbelievers challenge the efficacy of the later drug cocktails, HAART, is said to be the reason for describing as "AIDS denialists" Mbeki and other dissenters from the mainstream view (p. 82). But surely "AIDS denialist" means someone who denies that AIDS exists, not someone who questions the efficacy of drugs or that HIV causes AIDS. The claim by Nattrass and Nathan Geffen—another non-scientist activist— that providing HAART was economically feasible displays considerable political naiveté given that "some increase in tax revenue was probably needed" (p. 96). Chapter 5 concludes that Mbeki probably questioned HIV/AIDS because the issues "resonated with him intellectually" (p. 102). In other words he found the mainstream evidence and interpretations less than convincing. Hardly a reason to criticize him: Why were the mainstream proponents on the panel he organized unable to convince him?

According to Nattrass, "Buying into the world of AIDS denialism is seemingly empowering and exciting" (p. 109). No. It is intensely frustrating, for one thing because the mainstream simply ignores the evidence we point to, and for another thing because there is no overall organization and various individuals and groups are at loggerheads with one another over scientific points and over strategy and tactics for combating the mainstream. To flesh out her scenario, in Chapter 6 Nattrass invents stereotypes:

> ➤ the hero scientist (think Galileo);
> ➤ the cultropreneur who promotes alternative treatments—and Nattrass has fewer than five genuine examples, so this is entirely uncharacteristic of HIV/AIDS disbelievers;
> ➤ living icons: the "HIV-positive" individuals who are healthy without antiretroviral drugs;
> ➤ praise-singers: "a sizeable group of sympathetic journalists" (p. 127). Nonsense. The media coverage is extraordinarily rare, that treats disbelievers as other than misguided cranks.

Chapter 7 is about the censorship of the journal *Medical Hypotheses* by Elsevier. It is utterly misleading about peer review and boundary work (Bauer 2011) as well as about the censorship story. In particular, Elsevier did not act "quickly" by commissioning an "expert panel": Rather, Vice-President Glen Campbell withdrew the articles at issue within days of receiving protests from "HIV scientists," and Elsevier then spent many months looking for ways to justify that precipitate action, taken without consulting the articles' authors or the Journal's editor or editorial board (Bauer 2012). Chapter 8, in the context of "the struggle for evidence-based medicine," asserts a similarity between Duesberg's situation and that of Andrew Wakefield, a similarity that simply doesn't exist. Nattrass's lack of sophistication is illustrated when she sees something sinister in the term "wellness" which is allegedly supplanting "health" and she says "alternative and complementary" is what used to be called pseudo-science, chicanery, quackery (p. 155); why then did the National Institutes of Health establish a National Center for Complementary and Alternative Medicine (http://www.nccam.nih.gov)?

This book is hardly a credit to Columbia University Press, given the substantive demerits set out here. There are also unseemly *ad hominem* remarks (e.g., p. 113 about Nobelist Mullis). In addition, there are rather too many typos and strange expressions such as "foistered" and "heroizing."

HENRY H. BAUER

Professor Emeritus of Chemistry & Science Studies, Dean Emeritus of Arts and Sciences
Virginia Polytechnic Institute and State University
hhbauer@vt.edu; www.henryhbauer.homestead.com

References

Baker, C. (2010). Sticks and Stones. http://www.omsj.org/blogs/sticks-and-stones
Barber, B. (1961). Resistance by scientists to scientific discovery. *Science, 134*, 596–602.
Bauer, H. H. (2003). The progress of science and implications for science studies and for science policy. *Perspectives on Science, 11*(2), 236–278.

Bauer, H. H. (2007a). *The Origin, Persistence and Failings of HIV/AIDS Theory*. Jefferson, NC: McFarland.

Bauer, H. H. (2007b). More HIV, Less Infection: The Breastfeeding Conundrum. http://hivskeptic. wordpress.com/2007/11/21/more-hiv-less-infection-the-breastfeeding-conundrum

Bauer, H. H. (2009a). Confession of an AIDS Denialist—How I Became a Crank Because We're Being Lied to about HIV/AIDS. In Russ Kick, Editor, *You Are STILL Being Lied To: The REMIXED Disinformation Guide to Media Distortion, Historical Whitewashes and Cultural Myths*, New York: The Disinformation Company, pp. 278–282.

Bauer, H. H. (2009b). The Family of Rethinking AIDS. http://hivskeptic.wordpress.com/2009/11/15/ the-family-of-rethinking-aids

Bauer, H. H. (2011). Nattrass, Defending the Boundaries. http://www.respond2articles.com/SHIL/ forums/thread/861.aspx

Bauer, H. H. (2012). Chapter 3: A Public Act of Censorship: Elsevier and Medical Hypotheses. In *Dogmatism in Science and Medicine: How Dominant Theories Monopolize Research and Stifle the Search for Truth*. McFarland. pp. 74–91.

Crewdson, J. (2002). *Science Fictions: A Scientific Mystery, a Massive Cover-Up, and the Dark Legacy of Robert Gallo*. Boston: Little, Brown.

Duesberg, P. H., Mandrioli, D., McCormack, A., Nicholson, J. M., Rasnick, D., Fiala, C., Koehnlein, C., Bauer, H. H., & Ruggiero, M. (2011). AIDS since 1984: No evidence for a new, viral epidemic—not even in Africa. *Italian Journal of Anatomy and Embryology, 116*, 73–92.

Farber, C. (2006). Out of Control: AIDS and the Corruption of Medical Science. *Harper's Magazine*, March, 37–52.

Hook, E. B. (Editor) (2002). *Prematurity in Scientific Discovery: On Resistance and Neglect*. Berkeley, CA: University of California Press.

Kuhn, T. S. (1962/1970). *The Structure of Scientific Revolutions*. Chicago: University of Chicago Press.

Stent, G. (1972). Prematurity and uniqueness in scientific discovery. *Scientific American*, December, 84–93.

Weiss, S. H., & Cowan, E. P. (2004). Chapter 8: Laboratory Detection of Human Retroviral Infection. In Gary P. Wormser, Editor, *AIDS and Other Manifestations of HIV Infection* (fourth edition).

Journal of Scientific Exploration, Vol. 26, No. 4, pp. 892–894, 2012   0892-3310/12

The Origin of Everything, via Universal Selection, or the Preservation of Favored Systems in Contention for Existence by D. B. Kelley. Newbury, Ohio: Woodhollow Press, 2013. 339 pp. $32.95 (paperback). ISBN 978-0985462505.

The great problem in writing a theory of everything is that it may turn out to be a theory of nothing. Here is how it works. If you develop a theory that explains only some small, simple Thing, then the theory is very strong. It is precise, understandable, and it always works. As you expand the theory to encompass another Thing, it becomes weaker. It may still be precise and understandable, but it is now more complicated, and because it involves two things rather than one, it starts to become conditional. This means that in order for it to work with regard to the second Thing, we may have to take into account something about the first Thing. And so it goes. As the theory covers more and more Things, it becomes less precise, less understandable, and parts need to be added on in order cover multiple contingencies.

People who go down this path inevitably come to a critical point, at which their theory has become so precisely complicated that it can barely be understood, and gargantuan efforts become necessary to make it work at all. To resolve this impasse, they have a terminological epiphany. They suddenly find a language (that is, an arcane vocabulary) which confers the appearance of simplicity to the morass they have created. Precision is replaced by undefined terms and relationships, described in forceful but impenetrable prose. Argumentation becomes not an activity of rational thought, but a magical experience in which a swirling array of undefined concepts assembles itself in just the right way to come to whatever conclusion is desired. In the mouth of its inventor the theory does indeed explain everything, but in the ears of the audience it explains nothing.

I do not know if this process was the origin of *The Origin of Everything*, but there are some telltale signs. Kelley regards that everything from atoms to the universe itself comprises "systems." There is a chapter titled Defining Systems, which talks about systems without ever defining them. Evidently "system" is another word for "thing." Kelley's systems have behaviors (which he calls "behavioralisms") that are in some way motivated by intents. Thus they choose to cooperate with each other, or to compete with each other, or else they come to some conciliatory compromise between these extremes. There is, in fact, a rather large literature on the concept of "system" and descriptions of the behavior of natural systems, virtually none of which is used or referenced here.

Systems that exist do so because they are "stable" (another undefined term). If they are not stable, they are "weak." They are encouraged to become stable by a universal process of Behavioral Selection, another chapter title in which the topic under discussion is neither defined nor explained. Here is a sample:

> [selection] presents itself as one of the most fundamental laws in all of Nature, not only because it's ultimately responsible for stability and change, but because it achieves them by forever encouraging behavior that is both efficient and productive. It does so by favoring behavior that is good relative to the assemblages at hand. It thereby instills conduct that is respective not only to the phenomenon, but to those systems both superior and subordinate to its own. It will thus be clear that it forces all ensembles into maximum equilibrium, as it is often the path of least resistance for everything and everyone involved that leads to the greatest measure of relative good.

If this passage speaks to you, and you think it provides a definition or some kind of insight, then you are among the audience for this book. If, on the other hand, you find that it is dominated by bald assertion and looping repetition (in addition to poor writing), then you are not. I did not select this passage because it is unusual; in fact, it is a very good representation of both the thinking and style of the entire book.

Kelley does not shrink from making immodest comparisons between his work and that of Charles Darwin, in which Darwin is clearly the lesser intellect. Not only the title but also the foundation of Kelley's thesis are grounded in Darwin; indeed, he refers to his ideas as "universal Darwinism." Darwin is taken to task, however, for failing to appreciate that his ideas were not limited to the world of biology, which he carefully and systematically investigated for some thirty years before publishing, but are in fact ubiquitous truths, finally revealed by Kelley. The reader may well wonder, if Darwin failed where Kelley has succeeded, why Kelley's book is not filled with the careful working out of the consequences of a persistent program of intelligent and directed observation of the natural world. Why does Kelley think that vague and disconnected references to concepts in physics, joined with homey and familiar analogies from daily life, represent an advance over the methodological clarity and modesty of Darwin? Dare we think that *The Origin of Everything* is a sign of how far the practice of science has fallen in the 150 years since *On the Origin of Species*?

It seems to me possible that Kelley's whole selection idea is related to a common misinterpretation of Darwin. The phrase at issue is "survival

of the fittest," which can probably be parsed as "those individuals that survive are the ones that have the greatest capacity to survive." Stated this way, the principle is as true as it is barren. Saying that systems are selected (somehow) because they are "stable" might be insightful if we had a definition of "stable," but without such a definition the assertion might be parsed as "systems are selected because they are favored for selection." In fact, Darwin had perfectly good models for selection, involving predator–prey relationships, and competition for scarce resources, such as food. The only inescapably logical part of his reasoning was that an animal had to survive to a certain age in order to reproduce. By the mechanism of inheritance, reproduction implemented the last step of the selection process. For Kelley, systems reproduce themselves through an expression of their endlessly cyclical existence. Even more mysteriously, we read that

> Although selection thus appears as one of the most influential principles in Nature, however ironic, it involves a process that in some regards does not exist. . . . Seemingly contradictory to everything that we have learned thus far, sometimes neither Nature nor any other system makes any discriminatory selection at all.

I found the book to be increasingly difficult to read, as it wandered among a variety of topics, mentioning important works by famous authors, evidently hoping that propinquity alone would associate these great ideas with universal Darwinism. The insistent, unsupported repetition of the book's essential themes gives one the same sense as reading Voltaire's *Candide*, in which Dr. Pangloss insists on finding that we live in the best of all possible worlds despite the mounting contradictory evidence. Or perhaps Rudyard Kipling's *Just-So Stories*, in which increasingly confabulated explanations are found for every observable fact.

This book was published by Woodhollow Press in Ohio (Kelly evidently lives in Cleveland). I could not find a webpage for this press, and none of the book databases (including Amazon) listed any publications. In fact, the only web presence of Woodhollow I could locate was (multiple cites of) a laudatory press release about Kelley's book. In the modern era of publishing it is probably worthwhile to spend some effort on ascertaining the reputation and record of the publisher as part of the decision to acquire a book.

MIKEL AICKIN
National College of Natural Medicine, and University of Arizona

Journal of Scientific Exploration, Vol. 26, No. 4, pp. 895–906, 2012   0892-3310/12

Adventures in the Orgasmatron: How the Sexual Revolution Came to America by Christopher Turner. New York: Farrar, Straus, and Giroux, 2011. 532 pp. $35.00. ISBN 978-0374100940.

Christopher Turner writes for *Cabinet* magazine, a British quarterly that publishes articles about many facets of society, culture, science, and what have you, some of them allegedly rendered in a scholarly way.

The present book, Turner's first, is, I believe, intended to be scholarly in that Turner sets out in some detail the historical background for what he perceives to have been the sexual revolution. In it, he provides interesting, reasonably well-written information about political and cultural conditions in Europe and the United States in the early and middle part of the 20th century, when the modern sexual revolution germinated and came into full force.

However, it is difficult to decide whether the central theme of the book is the sexual revolution per se or the life and work of Wilhelm Reich. Of course, the two are inextricably bound together, as it was Reich who laid the scientific foundation for understanding sexuality in depth and for actively educating professionals and the masses about these facts and their significance for physical and emotional health and societal functioning. It is Reich's book *The Sexual Revolution* (1945 in English) that originally documented this shift in societal mores. Other pioneers in modern times who worked toward changing our sexual mores—Freud, other psychoanalysts, Kinsey, Marcuse, and Perls—also are mentioned and their work reasonably described, but they are given short shrift compared to Reich. It is here, however, that scholarly objectivity disappears.

Before offering the reader my critique of Turner's book, I think it well worthwhile to offer a very brief synopsis of Reich's work since some readers of this *Journal* may not be familiar with its scope. Wilhelm Reich was born in Austria in 1897 and, as a physician, became a practicing psychoanalyst in Vienna in 1919 at the age of 22. His elders and peers, including Freud, recognized his brilliance, and his seminal work on the analysis of character was accepted in large part by the psychoanalytic community as a major modification of the standard technique of free association. In addition to character analysis, Reich's major discovery during the psychoanalytic phase of his work was the clinical finding, made through detailed questioning of the sexual practices of his patients, that the majority of them by far were sexually impotent, that is, incapable of true gratification from the sexual act. Women were anesthetic or focused exclusively on clitoral climaxes rather than vaginal orgasm; men were often erectively impotent or suffered from premature ejaculation. When they were potent, too often they were incapable of surrendering emotionally to their partner. In all cases there was a great fear of giving in to orgastic pleasure. When through the character analytic process they overcame this fear, they rapidly lost their neurotic symptoms. Following Freud's original thinking, Reich thought that this fact indicated that there was an "economic" factor in neurosis: If the patient became capable of discharging stored excessive quantities of libido, neurotic fixations lost their energy source and, as a result, their power to influence the individual. This quantitative energetic factor became a cardinal element in all his following work.

Extensive deep investigations of social practices convinced Reich that neurosis might be treated on a mass level through education about healthy sexuality. Therefore, through a temporary alliance with the Communist Party in Berlin, where he was practicing at the time, he organized clinics and rallies where people, including adolescents, could receive information about healthy sexual practices including contraception. The rallies were attended by thousands eager for scientific information about sexuality and life.

In Reich's practice of psychoanalysis this meant that the goal became increasing the capacity of patients to surrender to their deepest impulses. When this could be done in the sexual embrace with a loved partner, bound energy would be adequately discharged and neurosis cured or prevented. This could not be done, however, simply by wish or command, because patients consciously and unconsciously resisted such surrender. The resistances took the form of character attitudes on both psychological and somatic levels. The psychological attitudes, the "character armoring" described by Reich, were rather well-known, but Reich also discovered that the psychological attitudes were anchored in chronic muscular tensions (muscular armoring). Complementing the analysis of the patient's character, dissolution of the muscular armoring was now utilized as a means of helping the patient overcome his/her resistances. This innovation became the basis of what is now known as "body work" in therapy, utilized by thousands of practitioners independent of those studying and utilizing Reich's methods on a more formal basis.

In the process of conducting "vegetotherapy," as Reich initially called the method, he noted that with the dissolution of the patient's armoring strong clonisms and pulsations appeared along with "electric currents" that patients described coursing through their body. Following a bioelectric theory of life extant at the time, Reich postulated that the libidinal energy was electrical in nature and that the fundamental life process was its spontaneous pulsation, its rhythmic expansion and contraction. In the 1930s Reich studied this phenomenon experimentally with a DC millivoltmeter of his own design. Readings on subjects' bodies, when in acute emotional states of anxiety, anger, or pleasure, confirmed Reich's concept of spontaneous organismic pulsation. Certain of the objective findings did not, however, fit an electrical concept. This set Reich on the path to the discovery of what he later called "orgone energy."

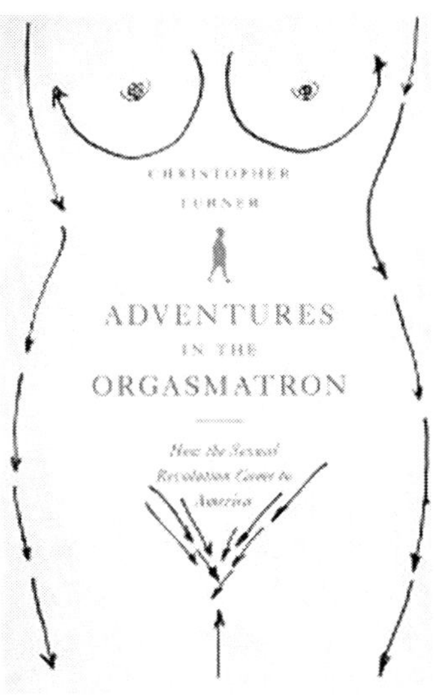

By examining boiled foodstuffs, a natural source of life energy, Reich found under sterile conditions using high-magnification microscopy that, no matter what the original source, all foods broke down into microscopic vesicles, which Reich called "bions." The bions consisted only of a membrane and some inner fluid, water. They glowed blue and moved from place to place in the microscope field. Bions could also be secured from non-organic sources such as carbon, iron filings, and ocean sand by heating them to incandescence and placing them in sterile nutrient media. Remarkably, they divided. Cultures of bions from ocean sand seemed to luminate strongly and the laboratory containing them showed anomalous effects: light phenomena such as lightning-like tiny rays, a blue glow in the air, the magnetization of metallic instruments, and light impressions on closed film cassettes without exposure to light. There were also strong biological effects such as "sunburned" skin exposed to the bions, and conjunctivitis in the eye Reich used to view bions in the microscope. Fearing some form of nuclear radiation, Reich had the bion cultures tested by a radiation specialist at a nearby Oslo hospital. Nuclear radiation was ruled out as were all possible other known forms of radiation. Reich was forced to conclude that he had discovered a previously unknown kind of radiation, which he called "orgone" because he discovered it in the course of his study of the sexual orgasm and because it was absorbed by organic materials. Metals attracted and repelled it as revealed by electroscopic investigations.

In order to examine the luminations more carefully, Reich built an enclosure consisting of non-metallic walls and an inner metal lining. He reasoned that, with the bion cultures within the enclosure, the orgone radiation would be concentrated inside. The light phenomena were more readily seen, but to Reich's surprise the phenomena remained after removing the bions and even after washing down the inner metal lining. Even building a new enclosure without placing bions within it showed the same luminations. It was clear to Reich that the enclosure was somehow concentrating a radiation that existed in the atmosphere. Reich named the enclosure the "orgone energy accumulator" (ORAC). He demonstrated many powerful biological effects, including prolongation of the life of mice with cancerous tumors as well as anomalous physical effects within the enclosure such as a spontaneous elevation of temperature and an anomalous prolongation of discharge times of statically charged electroscopes. Later he found that Geiger-Mueller counters could also detect the orgone. In the next decade, Reich was able to experimentally demonstrate a motor force by hooking specially prepared devices to accumulators; anomalous interactions between nuclear radiation sources and concentrated orgone energy; and weather modification by a device that could alter the distribution of orgone energy in the atmosphere.

All of these experiments, including aspects of the nuclear radiation experiment, have been independently replicated by scientifically competent investigators including myself and have been published in a variety of scientific journals.

Wilhelm Reich died in 1957. Since then four books have been published on Reich and his science, orgonomy. These are *Wilhelm Reich and Orgonomy*, written by Reich's Norwegian student Ola Raknes (Raknes 1970), David Boadella's *Wilhelm Reich, the Evolution of His Work* (Boadella 1973), Elsworth F. Baker M.D.'s *Man in the Trap*, about psychiatric orgone therapy (Baker 1967), and Myron Sharaf's *Fury on Earth*, a comprehensive biography of Reich (Sharaf 1983). All presented Reich and his work accurately and sympathetically, although limited in scope by their individual areas of expertise. Now we have Christopher Turner's *Adventures in the Orgasmatron: How the Sexual Revolution Came to America*. Within the context of what was happening in the world at that time, Turner, herein, reports on most of the scientific issues described above.

In Woody Allen's film *Sleeper* there is a complex-looking cabinet-like device called the "orgasmatron." Its function is obvious from its name. Turner would like us to believe that the device bears a reasonable similarity to Reich's orgone energy accumulator: Therefore, one should take neither of them seriously. And that sets the tone for the entire book. Unlike Sharaf, Turner is neither accurate in his reportage nor even-handed. Indeed, there are so many factual errors in his book that one is inclined to say that Turner "isn't even wrong," as Nils Bohr said of a student in advanced physics. Turner gives us the references, the footnotes, the conversations with those who knew Reich, and so on in an effort to persuade us that this is a scholarly work, but to no avail; anyone checking up on Turner's quotes and allegations will find innuendo, diversions, half-truths, and outright lies all designed to denigrate Reich and make him appear as a crazy fraud.

The deception begins in the Introduction, where Turner would have us believe that Reich came to the U.S. in order to spread his findings about sex and politics. Factually, with the rise of Nazism, it was no longer tenable for Reich to stay in Norway where he had made his major biological discoveries. Reich came to the U.S. on the invitation of Theodore Wolfe, M.D., a Professor of psychiatry at Columbia University, to lecture on his medical discoveries at the New School for Social Research in New York City. Still in the Introduction, the ORAC is described as, "a box in which . . . his [Reich's] ideas came almost prepackaged . . . an almost magical device that could improve its users' orgastic potency . . . and their mental health," both of which claims Reich categorically denied, although they became the mantra used by all who wanted to attack him, and which, conversely, identified his attackers. A bit further on Turner alleges that, "People sat in the orgone box hoping to dissolve the toxic danger of conformity." This is pure nonsense and simply unsupported opinion on Turner's part. At the end of the Introduction we find out Turner's M.O.: The ORAC is to be seen as "a prism through which to look at the conflicts and controversies of that era [the era of the sexual revolution coming to America]." For those of us who really know about Reich's work, it soon becomes clear that the eye of Turner is peering cockeyed through this prism.

Turner admits that Reich made seminal innovations in psychoanalytic method with his discovery of character analysis, a method of analyzing how a patient presented himself rather than the content of his associations, a method accepted by psychoanalysts which revolutionized psychoanalytic technique. However, Turner denigrates all the other discoveries that complemented character analysis. These were the orgasm theory, the rooting of psychic disturbances in chronic muscular tension (muscular armoring), and the social consequences of successful character analysis. The criticism is not rendered impartially but contemptuously, utilizing sources who knew little or nothing about Reich or his work. Often the source simply dismisses the man or his work as "crazy" or "ridiculous." For example, after reasonably describing Reich's original findings about the nature of the impulsive character, made while Reich was first assistant at the Vienna psychoanalytic clinic, Turner blithely labels Reich as having that same diagnosis based on something that Dr. Elsworth Baker, an experienced psychiatrist and therapist and Reich's most able American student, wrote about him. Such a diagnosis implies instability, extreme emotional lability, secretiveness, and a tendency to sociopathic behavior. Baker did write that Reich was "impulsive" (Baker 1976:182). But there is a world of difference between having a psychological trait, such as impulsivity, which anyone can have, and being someone who has that trait rule his life, as is so for a character type such as the impulsive character. It is obvious that Turner has little understanding of Reich's characterology, declaring Reich to be an impulsive character, as just mentioned, or as "schizophrenic," or as "manic-depressive," depending upon whom he wishes to quote at the time. Borrowing most heavily from those who became enemies of Reich as he traced a meteoric path through science and society, Turner never presents substantive information that would confirm any of these diagnoses.

Those who worked with Reich knew of his intensity, that he marched to a different drummer, and that he did not tolerate fools, but generally his co-workers recognized his genius, kindness, and capacity for deep emotional and interpersonal contact. These are not the qualities of someone with severe emotional disturbances. Those who could not keep up with Reich on the level at which he functioned either dropped out with grace, or too often fell by the wayside, furious at Reich for abandoning them. Turning on Reich, they accused him of their own shortcomings, trying to tear him down to their level. Reich called this "the emotional plague."

This is not to say that in his later years Reich was not at times emotionally agitated, especially when having drunk too much. It is not surprising, when considering the great amount of disappointment and calumny Reich received from many of his contemporaries after opening his heart to them and to the world to the great extent that he did, that he developed some extreme, defensive postures. But as Baker put it, describing Reich, "impulsive, but insane, never."

In a book that purports to be an accurate historical document, I was surprised to find more than 30 factual errors about scientific matters alone. Many of them were the same old canards about the orgone energy accumulator (ORAC), for example: that it was a Faraday cage (it is not grounded, as is a Faraday cage); that it was an atomic shelter in reverse . . . where the radiation could be "contained and neutralized" (never one of Reich's concepts); that one could increase one's orgastic potency and have orgasms by sitting inside an accumulator (not so, as Turner himself knows by quoting Reich, "I wished it did, but it does not"); that the ORAC ". . . dissolves the toxic dangers of conformity" (not so—the ORAC has nothing to do with social change, being strictly a physical device); that the ORAC could cure cancer (never claimed by Reich, despite experimental evidence that it could prolong life in mice with spontaneous tumors); that the metal lining of the ORAC stopped orgone energy from escaping from the enclosure (not true—metals first attract and then repel orgone, as demonstrated experimentally); that Einstein proved that an observed, apparently spontaneous elevation of temperature within the ORAC could be explained by convection (not so, as Einstein failed to control for his own experimental refutation of the observed phenomenon); that Reich claimed that his laboratory was radioactive after placing one mgm of radium within an ORAC (not so, although an anomalous elevation of background counts was measured even when the radium had been removed 1/4 mile away to a lead container within a steel safe—this led Reich to conclude that the orgone had been excited to a new state of functioning); that Reich's invention, the orgonoscope, a device for visualizing orgone energy, could move waves (not so, as the orgonoscope is a closed, inches-long tube, whereas what Reich described as disturbing the surface of a lake was a hollow metal tube several feet long); and so forth.

Regarding Reich's biological discoveries, it is clear that Turner either had not carefully researched Reich's publications or chooses intentionally to misrepresent them. For example, in discussing Reich's discovery of the "bions," microsopic vesicles that develop spontaneously in disintegrating organic and inorganic matter, Turner writes as if Reich were claiming he had discovered particles that were alive and originated "de novo," as it were. In fact, Reich was careful to describe the bions as not arising de novo but as being transitional states between the non-living and the living.

Turner reports on the FDA's scientific case against Reich. While preparing to indict him for transporting a "fraudulent" medical device—the ORAC—across state lines, various of Reich's experiments were allegedly replicated by scientists and physicians at different laboratories. Turner describes some of their results, all negative, but also includes, without comment, the scientists' attitude toward the very work they were asked to do. A three-man committee on mathematical biophysics at the University of Chicago found the accumulator to be "a gigantic hoax with no scientific basis" on purely theoretical grounds. One of them said, "The material is beneath any refutation." A physician testing for basic physiological reactions stated, "It was very difficult for me to bring myself to take the time to prepare this report because . . . this quackery is of such a fantastic nature that it seems hardly worthwhile to refute the ridiculous claims of its proponents." So much for scientific objectivity and openness of mind!

A physician tested a variation of the ORAC on a trichomonas infection of the vagina and found in one case that the infection cleared up immediately after treatment. This was written off as due to a strictly mechanical effect of introducing the device into the vagina. According to Turner, when a physician consulting for the FDA was presented with Reich's finding that red blood cells from cancer patients develop "spikes" in physiological saline solution, he said the spikes were ". . . the natural crenellation [sic] of red blood cells." Anyone who has faithfully performed this blood test using Reich's strict protocol can easily differentiate naturally crenating (scalloped edge) red blood cells from the spiked cells described by Reich. Since physicians are familiar with *crenation*, the use of the term *crenellation*, which refers to the embattlements of forts, not red blood cell disintegration, must have been Turner's error. Furthermore, the spikes rarely develop "naturally," but only where there is an energy-deficient chronic illness such as cancer. Obviously Turner doesn't know what he is writing about here.

I was working at the Jackson Laboratory in Bar Harbor, Maine, the summer that the FDA granted the lab funds to conduct a test of the ORAC on cancer mice. On speaking to the assistant who was conducting the experiment, I inquired and found out that the treated mice were dying significantly faster than the controls. On Reich's suggestion, I found out that an X-ray machine was in close proximity to the laboratory where the studies were done. Reich had previously found that the presence of high-frequency electromagnetic radiation induced a negative, disturbing effect on the radiation within the ORAC and he asked me to explain this to Dr. William Murray, the scientist in charge of conducting the study. I did so, suggesting to Murray that he read the literature where this effect is described. Murray told me that "I won't do that because I don't want to prejudice myself while running the experiment."

After detailing all the negative reactions to the ORAC, you would think that Turner would make an effort to balance his reportage by documenting some positive comments. Nary a one! And he had ample access to literature on experimental work using the ORAC published by James DeMeo, Ph.D., Dr. Stefan Muschenich who found an anomalous elevation in temperature in subjects using the ORAC compared to suitable controls, and myself and others who replicated Reich's cancer experiments on mice. Nor does Turner refer to the detailed analysis of the FDA's scientific experiments published by Dr. Courtney Baker (writing under the pseudonym C. F. Rosenblum) and myself (Blasband & Rosenblum 1972/1973). Obtaining them under the Freedom of Information Act, we found the FDA studies shoddy work at best, something that would never be published in any self-respecting scientific journal. We even found some results that confirmed Reich's findings, but which were not mentioned in the conclusions of the articles. In most cases, Reich's strict protocols were assiduously avoided.

It is often difficult to differentiate the blockheadedness of the FDA inspectors and scientists from Turner's own pathetic investigation, reportage, and opining. For example, Turner writes that FDA inspectors, when first visiting Reich's laboratory, carried radioactivity-monitoring film badges and dosimeters, allegedly because Reich had written in *The Oranur Experiment* (Reich 1951) that his premises were "dangerously radioactive." Reich never wrote this. What he did write was that small amounts of radium placed within an ORAC triggered off a field reaction in the orgone, causing G–M counters to output anomalous counts. In fact, as reported by Reich after the initial reaction, the 1 mgm of radium used in the experiment anomalously lost much of its radioactive quality, as measured by electroscopic discharge (Reich 1951).

With respect to all of Reich's psychological, biological, and physical research, I could find few areas in which Turner is not confused. For example, he describes Reich as finally embracing Freud's death instinct. In fact, what Reich had discovered was a toxic state of orgone energy ("DOR") which had life-negative qualities. Reich ventured that Freud's perception of a wish for death that he could see in some people had its biophysical basis in DOR. But this is a far cry from embracing the concept that people had an instinct to embrace death, as Freud alleged, in describing masochists who defied recovery despite extensive psychoanalysis. Nor do we find Turner giving Reich credit for solving the problem of masochism with his character-analytic technique.

Turner describes Reich as having "assumed the *mythic* status of rainmaker" [italics mine] in conducting his weather control experiments. He then goes on to describe how Reich actually did make it rain according to Reich's son Peter. No comment, however, by Turner about this apparent contradiction. Perhaps it wasn't a myth?

Nor is there any mention of the well-documented orgonomic weather work over a thirty-year span, published by myself in the *Journal of Orgonomy*, nor of Dr. James DeMeo's well-documented weather work in the U.S. and abroad, as well as DeMeo receiving his master's degree (thesis published) at the University of Kansas for his controlled study of the use of the Reich weather apparatus, the "cloudbuster," in generating rainstorms.

Turner's use of the term *mythic*, above, is typical of how in other contexts Turner denigrates Reich's work. Turner describes the ORAC as being similar to a 19th-century wooden cage named the "Utica Crib." Why? Because, Turner writes, Otto Fenichel, a psychoanalytic associate of Reich's in Berlin, allegedly circulated the rumor that Reich had been hospitalized in the Utica State Hospital!! In another context Turner implies that Reich is schizophrenic. How come? Because, as Turner states, Reich understood how schizophrenics functioned. So it follows that, in Turner's loosely associative mind, to Reich the ORAC must be little more than his very own "Influencing Machine," a fantasied device described by a schizophrenic patient of the psychoanalyst Tausk.

Turner interviewed or obtained quotes from many people who had known Reich, including his earlier psychoanalytic colleagues, his daughter Lore, and his son Peter. Except for Reich's student Dr. Elsworth Baker and Peter, all of them opined that Reich was either schizophrenic or manic-depressive. There is no substantiation of these opinions: Indeed all of them are based upon the fact that Reich talked and wrote about "energy" in people and in the atmosphere. Since such concepts were unknown at that time (1930–1940), except for the mainstream concepts of physical energies, it was simply accepted that Reich was crazy and hallucinating. Turner goes with the opinions: When Reich sits in the ORAC he sees fog-like formations, bluish dots, lines of light, and violet light phenomena apparently emanating from the walls. Turner understands this as Reich having "hallucinations." Reich, the observing scientist, spent long periods of time in the ORAC to substantiate his original subjective impressions. This is understood by Turner as ". . . Reich being locked in his iron cage (as) testament to his increasing alienation." Indeed, as Dr. DeMeo, myself, and others have found, anyone sitting in an orgone room for more than a half hour on a dry, sunny day will see just what Reich saw, without being prepped to do so.

When it comes to presenting Reich's views on sexuality, Turner does no better than most who have taken on this subject (Blasband 2006). Initially, at least, Turner appears to get it right when he quotes Reich: "It is not just to fuck, you understand, not the embrace in itself, not the intercourse. It is the real emotional experience of the loss of your ego, of your whole spiritual self." Although Turner claims familiarity with Reich's work, it appears that he doesn't seem to really understand what Reich is talking about. Both subtly and not so subtly, Turner derides Reich's concept of "orgastic potency," and gives testimony by psychoanalysts and other prominent individuals refuting it. An examination of the testimony reveals, however, a failure to distinguish between primary and secondary sexuality. As Reich discovered clinically, primary drives are impulses moving out toward the world unimpeded by armoring, from the deepest part of the self—love fused with eroticism. Secondary drives are primary impulses that become distorted as they are expressed through the psychic and muscular armoring—loveless sex, sadism, pornography. In Western society, sadly, the primary drives are hardly known; the secondary drives are considered to be the norm, one's "nature."

For example, Turner, searching to bolster his view that Reich was some kind of sexual nut, quotes James Baldwin: "There are no formulas for the improvement of the private or any other life—certainly not the formula of more and better orgasms. . . . The people I had been raised among had orgasms all the time, and still chopped each other with razors on Saturday nights" (Baldwin 1961). And Turner would have us believe that Reich thought that the cure of neurosis could be effected simply by having patients have sexual intercourse. Of course that is nonsense, since most intercourse lacks the gratification in energetic discharge that permits the establishment of a healthy "energy economy." The reason? Emotional armoring against pleasure, established originally in childhood and perpetuated by a sex-negative society. It is not a matter of "how many times one can do it," but of the quality of satisfaction and gratification in the sexual embrace. Sex without love can never result in this. Turner's further evidence against Reich's thesis? The sexual libertarianism of the Nazis and the fact that this did not lead to political freedom!! Again, no serious distinction between "fucking" and sex with love. According to Reich and the experience of many of his students and patients, the process of therapy can spontaneously establish primary sexuality and secure emotional health.

Turner's utterly abysmal reportage is well-illustrated in his attempts to make Reich out as a sexual pervert. According to Turner, Lore Reich Rubin, Reich's youngest daughter, told Turner that her father was a sexual pervert, at the very least a voyeur. "I wouldn't be surprised," she is quoted as saying, "if he molested my sister, though she would never admit that, I'm sure. . . . He was really a sex abuser, excuse me for saying it . . . I don't have any evidence, but I think he was." Here, Turner accuses Reich of sexual perversity and using as evidence total hearsay from Reich's daughter, who earlier in the interview, as Turner reports, states that *she didn't think there would have been anything wrong* [italics mine] with it if Reich had made sexual advances toward her when she was a child. If true, what an incredible statement!

Innuendo is heaped upon innuendo to paint the picture of Reich as a pervert, and Reichian therapists as therapeutic sadists, seducers of children, and rapists. Turner reports that Susannah Steig, the niece of the cartoonist William Steig—himself an ardent follower of Reich, ". . . tells of another Reichian therapist who allegedly repeatedly raped an eleven-year-old patient for months; apparently the unnamed analyst was later put into a mental institution." In summary, this book is muckraking from the bottom of the barrel; much alleged, little evidence. The popular press loves it. To date, at least nine book reviews extoll Turner's acumen and revelatory reportage, delighting in bringing Reich down. None view the book with a truly critical eye. How come? The answer lies in what Reich found to be true about others' reactions to his findings and himself and, indeed, to most pioneering scientists and thinkers who have disturbed man's emotional and cognitive equilibrium. Man has little capacity to tolerate the truth about himself: Excited by orgonomy and incapable of tolerating this excitation, the *Little Man* (Reich 1948), like Turner, attacks, quelling his inner disturbance by getting rid of the person who caused it. Put down Reich, try to kill him and his work, no matter what nefarious means are used, so the Turners of the world can breathe easier. Reich called this the Emotional Plague.

This is not a pejorative term. Reich described the emotional plague as a medical problem. One feels genuine sadness that such responses as Turner's make it more difficult for the light of geniuses and pioneers like Reich to shine into the dark corners of the world.

Acknowledgments

I am indebted to my former colleagues in the Cosmic Orgone Engineering (CORE) network, and to my dear friend, Allaudin Mathieu, for their editorial assistance.

RICHARD A. BLASBAND
Center for Functional Research, Sausalito, California

Note: A version of this review appeared in *Subtle Energies*, *21*(3), in 2010.

References

Baker, E. F. (1967). *Man in the Trap*. New York: The Macmillan Co.
Baker, E. F. (1976). My eleven years with Wilhelm Reich. *Journal of Orgonomy*, *10*(2) (November), 182.
Baldwin, J. (1961). The New Lost Generation. *Esquire*, July, 113.
Blasband, R. A., & Rosenblum, C. F. (1972/1973). An analysis of the U.S. FDA's scientific evidence against Wilhelm Reich. Part I: The biomedical evidence (Blasband), and Part II: The physical concepts (Rosenblum), *Journal of Orgonomy*, *6*(2), November 1972. Part III: The physical evidence (Rosenblum), *Journal of Orgonomy*, *7*(2), November 1973.
Blasband, R. A. (2006). Review: *Sex on the Couch: What Freud Still Has to Teach us about Sex and Gender* by Richard Boothby. *Journal of Scientific Exploration*, *20*(3), 479.
Boadella, D. (1973). *Wilhelm Reich, the Evolution of His Work*. New York: Laurel Editions.
Raknes, O. (1970). *Wilhelm Reich and Orgonomy*. New York: St. Martin's Press.
Reich, W. (1948). *Listen, Little Man*. New York: Orgone Institute Press.
Reich, W. (1951). The Oranur Experiment. *Orgone Energy Bulletin*, *3*(4), October 1951, 347.
Sharaf, M. (1983). *Fury on Earth*. New York: St. Martin's Press.

Journal of Scientific Exploration, Vol. 26, No. 4, pp. 907–908, 2012   0892-3310/12

Altering Consciousness: Multidisciplinary Perspectives Volumes 1 & 2 edited by Etzel Cardeña and Michael Winkelman. Santa Barbara, CA: Praeger, an imprint of ABC-CLIO, 2011. Volume 1, 401 pp. Volume 2, 399 pp. $124.95. ISBN 978-0313383083.

This two-volume set is an extraordinary accomplishment of scholarship. Cardeña and Winkelman have assembled a wide-ranging collection of chapters written from an array of multidisciplinary perspectives that nonetheless flows easily from start to finish. The depth and breadth of these two volumes is amazing. In addition to thoroughly covering classic topics such as hypnosis, meditation, shamanism, psychedelics, and sleep and dreams, there are chapters on the visual and performance arts, music, literature, philosophy, psychopathology, addiction, sex, human development, neurochemistry, and parapsychology.

Given the enormous range of topics and disciplinary domains, the editors are to be credited for facilitating conceptual continuity across chapters in terms of definitions and terminology. The editors also embed internal references linking chapters and subjects within and across the two volumes. The organization imposed on this diverse array of topics contributes to the smooth transitions and logical branching of ideas and observations. As a result, the volumes are well-integrated and far more readable than many edited books.

In Volume 1: History, Culture and Humanities, Cardeña and Winkelman contribute introductory chapters contextualizing altered states of consciousness and identifying the frames of reference within which they can be examined. The chapters are grouped into a series of "perspectives" from which to view altered consciousness. The first section, Historical Perspectives, details the evolution of Western thinking about altered states of consciousness capped by a reconceptualization of the last fifty years by Julie Beischel, Adam Rock, and Stanley Krippner. This section is followed by Cultural Perspectives, describing alternative understandings including Eastern, shamanistic, social, and technological approaches to altered states. The volume concludes with a rich set of chapters on the many and varied roles of altered consciousness in religion, the arts, and the humanities.

Volume 2, Biological and Psychological Perspectives, begins with a ten-chapter section on biological and pharmacological approaches to understanding altered consciousness. Sleep and dreams, dopamine, serotonin, psychedelic drugs, and sex are examined in detail. The chapter by

Andrea Blätter, Jörg Fachner, and Michael Winkelman provides an important perspective on the role of altered states in addictive behaviors that is missing from most scientific and social approaches to substance abuse. The final section, Psychological Perspectives, focuses on altered states as they are manifest in healing and spirituality and in psychopathology. The volume concludes with a chapter by David Luke summarizing and critiquing the limited research on altered states and paranormal phenomena.

There is also another reason why the chapters fit together so well. Implicitly or explicitly the authors conceptualize altered consciousness as manifest in the form of distinct mental states—trance states, meditation states, dream states, hypnotic states, drug states, vegetative states. Indeed the notion of the "stateness" of altered consciousness serves as a common denominator across the many disciplinary perspectives, levels of analysis, and sources of data represented in these volumes. The concept of altered consciousness as a distinct (or as Charles Tart would say a "discrete") mental state serves as an isomorphic construct that connects observations at one level of analysis with those at another. A *state* can be defined at multiple levels from the neurochemical to the behavioral, from the physiological to the psychological. So there is a conceptual convergence across chapters that links together discussions of music, religion, out-of-body experiences, yoga, dopamine, and a host of other phenomena.

A number of historical and scientific themes play out across the volumes as authors draw upon overlapping sources of data to make their particular case. William James is frequently quoted, as are the books by Charles Tart. In his autobiographical Preface, Extending our Knowledge of Consciousness, Tart recounts encounters at various points in his career with the unwillingness of mainstream science to examine the extraordinary— but methodologically messy—phenomena of altered consciousness. Other authors also allude to the seemingly blind resistance of conventional science to take this stuff seriously.

The question of what it takes to generate mainstream support for investigating the potential that may be contained within specific altered states of consciousness remains unanswered. But these two volumes advance the case by making accessible the rich and diverse domains of experience and knowledge that humankind possesses. They serve both as a comprehensive set of references and as a readable historical, cultural, and psychological narrative that summarizes the current state of the many arts and sciences of altered consciousness.

FRANK W. PUTNAM
University of North Carolina, Chapel Hill, North Carolina

Journal of Scientific Exploration, Vol. 26, No. 4, pp. 909–910, 2012   0892-3310/12

The Illustrated Encyclopedia of Native American Mounds and Earthworks by Gregory L. Little, illustrated by Dee Turman. Memphis, TN: Eagle Wing Books, 2008. 342 pp. $29.95. ISBN 978-0940829466.

Although the U.S. public is generally aware of the erections of central and eastern North America's pre-Columbian Mound-Builders, it does not widely appreciate how astonishingly numerous, extensive, and sophisticated (including lunar alignments) these earthworks were and how puzzling certain aspects of their histories and uses remain. Louisiana psychologist and investigator of the supernatural and the paranormal Gregory Little has written a very useful book of inventories, state by state, of hundreds (out of the original hundreds of thousands) of mounds and earthworks, many of which are illustrated with pictures from old U.S. Bureau of American Ethnology and other archaeological publications, supplemented by new visual reconstructions executed by illustrator/graphic designer Dee Turman. There are also a fair number of images of associated artifacts.

In the inventory, each mound/earthwork or complex covered is categorized by type (e.g., burial mound, effigy mound) and archaeological culture (e.g., Hopewell, Adena, Mississippian), and its exact location given (when rules and ethics permit), followed by a brief-to-lengthy description and discussion. The discussions are generally quite straightforward, but one encounters the occasional provocative observation—e.g., regarding Newark, Ohio's, Great Circle Earthworks: "Its size and basic layout is identical to England's [megalithic] Avebury site (except without the standing stones present at Avebury)" (p. 192). But part-Seneca Little does not think that these works derive from Old World models; in his brief Introduction, he emphasizes that "All we can say with certainty is that the ancestors of the present Native Americans [Indians] built the mounds" (p. 2). This is in contradistinction to the nineteenth-century notion that the mounds reflected arrivals from overseas on the part of Israelites and perhaps others.

The author outlines the success of Ohio State University economist J. Huston McCulloch in confirming the one-time existence of the now-destroyed *"Hannukkiah"* (sic)" earthworks near Milford, Ohio, whose plan was long thought to resemble the outline of a Hebrew lamp and menorah (p. 189). According to Little, two British archaeologists

Turn[ed] the map [of the "menorah"] upside down and the earthworks instantly became recognizable as a bird effigy. The bird's head is depicted, as well as a forked tail, and the wing feathers are flowing to the rear of the bird. (p. 190)

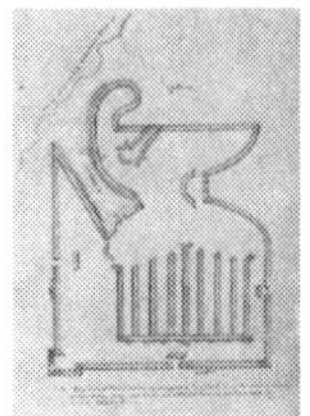One can see how this interpretation is possible, although it is not as obvious to me as it was to the archaeologists. Little does not mention McCulloch's work in attempting to interpret the intriguing Newark Holy Stones—several carved objects carrying Jewish symbolism and Hebrew writing, including a condensed Decalogue, said to have been unearthed in 1860 in a burial mound ten miles to the south of Newark, Ohio, and in a "well" next to Newark's Octagon earthwork—which mainline professionals perceive as fakes, but which McCulloch thinks genuine (http://www.econ.ohio-state.edu/jhm/arch/decalog.html).

Other inscribed tablets with Semitic texts have been reported from American mounds, the best candidate for demonstrable genuineness being that from East Tennessee's Bat Creek Mound—which Little does not include here. He does characterize such a tablet from West Virginia's Grave Creek Mound: "archaeologists universally accept the stone as a fake planted in the mound to gain attention for the commercial museum venture, which failed" (p. 261).

Until recently, with an age of about 3200 BC, Louisiana's Watson Break Mounds were considered to be the continent's oldest (p. 110); but Little points out that recent work in Lower Jackson Mound at the state's more-renowned Poverty Point site indicates a date there of about 3800 BC (p. 107)! This is well before the building of Egypt's pyramids.

It is worth noting that Louisiana State University-associated psychopharmacology/drug-rehab specialist Little, who holds an Ed.D. in counseling, is also the author/co-author of books in support of the ideas of the purported psychic and healer Edgar Cayce (1877–1945) and Cayce's Association for Research and Enlightenment (*Secrets of the Ancient World: Exploring the Insights of America's Most Well-Documented Psychic, Edgar Cayce*), on Atlantis (including the Bimini Road and other sites in the Bahamas), and Mu (*Edgar Cayce's Atlantis*; *Mound Builders: Edgar Cayce's Forgotten Record of Ancient America*; *The A.R.E.'s Search for Atlantis: The Ongoing Search for Edgar Cayce's Atlantis in the Bahamas*), and on UFO-abduction experiences (*Grand Illusions: The Spectral Reality Underlying Sexual UFO Abductions, Crashed Saucers, Afterlife Experiences, Sacred Ancient Ritual Sites, & Other Enigmas*; *People of the Web: What Indian Mounds, Ancient Rituals, and Stone Circles Tell Us about Modern UFO Abductions, Apparitions, and the Near Death Experience*); no such material is included in the scholarly book here reviewed.

STEPHEN C. JETT
Professor Emeritus of Geography and of Textiles and Clothing, University of California Davis
Editor, *Pre-Columbiana: A Journal of Long-Distance Contacts*, scjett@hotmail.com
Note: This review also appears in *Pre-Columbiana: A Journal of Long-Distance Contacts*.

Journal of Scientific Exploration, Vol. 26, No. 4, pp. 911–922, 2012   0892-3310/12

Metapsichica Moderna: Fenomeni Medianici e Problemi del Subcosciente by William Mackenzie. Rome: Libreria di Scienze e Lettere, 1923. 450 pp. ASIN B006MSD3YO.

The psychical researcher William Mackenzie (1877–1971) was born in Genoa of Scottish parents and spent almost his entire life in Italy, while formally remaining a British citizen. He studied at several Italian and European Universities, graduating in Biology and Philosophy, and in 1905 he established a marine biology laboratory in Quarto dei Mille. During that period he published his first important work, *Alle Fonti della Vita* (*To the Sources of Life*) (Mackenzie 1912a), and his speculative nature and his passion for the philosophy of biology are already evident. In that work, Mackenzie claims that materialism is insufficient to explain biological phenomena, and he drew the foundations of a neovitalistic theory which asserted the specificity of a biological realm ruled by an intelligent and psychic energy needed to explain the teleological complexity of living organisms.

Between 1912 and 1914 Mackenzie gained notoriety following the studies conducted with the founder of Psychosynthesis Roberto Assagioli (1888–1974) on the famous "calculating horses" of Elberfeld, which issued their responses to complicated mathematical calculations by rhythmically beating their hooves. The calculating horses attracted a large group of European scholars, including some of the most eminent Italian psychologists such as Father Agostino Gemelli (1878–1959) and Giulio Cesare Ferrari (1867–1932) (Zocchi no date). The studies were in the journal founded by the latter in 1905 in Bologna, *Rivista di Psicologia* (*Journal of Psychology*), a journal of central importance for the then-fledgling field of Psychology in Italy (Marhaba 2003:60ff). In 1912 Mackenzie published the paper *I Cavalli Pensanti di Elberfeld (The Thinking Horses of Elberfeld)* (Mackenzie 1912b) in that journal. Still in Germany and with Assagioli, Mackenzie studied Rolf the Mannheim's dog, another "thinking" animal supposedly with a strong metaphysical and mathematical intelligence. The synthesis of those years of research resulted in *Nuove Rivelazioni della Psiche Animale* (*New Revelations of the Animal Psyche*) (Mackenzie 1914), in which Mackenzie claimed that the phenomena were genuinely produced by the intelligence and will of animals.

William Mackenzie (1877–1971)

During World War I he was drafted by the Italian army in order to organize psychological assistance to the troops. During this time he published *Significato Bio-Filosofico della Guerra (The Bio-Philosophical Meaning of War)* (Mackenzie 1915), a philosophical analysis of war seen first of all as a phenomenon that cannot be eliminated because it is rooted in the biology of living beings.

During World War II, however, Mackenzie lived in exile in Switzerland: His British citizenship made him an enemy in the eyes of the Italian state, which expropriated all his material possessions. From 1939 to 1945 he still managed to earn a living teaching the Philosophy of Biology at the University of Geneva. One of the most successful courses which he held during those years was *Dix Étapes sur la Route de l'Esprit (Ten Steps on the Road of the Spirit)*, dedicated to the great spiritual masters of human history and their doctrines. In fact, he concluded that spirituality, read in the Jungian interpretation, was fundamental to explain the human being. The course was published several years later in French, Italian, and other languages under the title *Les Grandes Aventures Spirituelles (The Great Spiritual Adventures)* (Mackenzie 1967).

Undoubtedly, contact with the enigmatic phenomena of the animal psyche steered Mackenzie to psychical research. Over the years he became Honorary President of the *Società Italiana di Parapsicologia (Italian Society of Parapsychology)* of Rome, Member of the *Institut Métapsychique International* of Paris, and in 1955 he founded and directed a quarterly journal that lasted only two years, *Parapsicologia (Parapsychology)* published by the publishing house Fratelli Bocca, which collected many contributions from international researchers. Between 1915 and 1950 he experimented with various mediums such as Jan Guzyk (1875–1928) and Stephan Ossowiecki (1877–1944). However, it was the séances in which he participated in 1921 in Brussels, during which the alleged mediumistic entity Stasia manifested, that established him permanently in psychical research (Beverini 1988, Talamonti 1971).

The experiences with Stasia constitute one of the chapters of *Meta-*

psichica Moderna (*Modern Metapsychical Research*) (Mackenzie 1923), published in Rome in 1923 and quite recently reprinted under the title *Guida ai Fenomeni Medianici* (*Guide to Mediumistic Phenomena*) (Mackenzie 1988). In the Italy of the 20s the study of paranormal phenomena had already been ostracized by both psychologists and the academics, and was increasingly polarized between spiritualists and metapsychical researchers determined to oust the spiritistic hypothesis from the theoretical knowledge of the discipline; moreover, due to the aftermath of World War I, researchers were struggling to keep pace with the international community (Biondi 1988, Biondi & Tressoldi 2007). That is why *Metapsichica Moderna* (*MM*) can be seen not only as Mackenzie's greatest legacy to the field of psychic studies, but also his contribution as to the philosophy of biology, which is making metapsychical research a science and thus detaching it in a decisive manner from any form of spiritualism.

Contents

Mackenzie wrote *MM* to spread the view of metapsychical research he learned from Morselli (1908) and Richet (1922), which is a positive, factual, and empirical metapsychical research, free as possible from early hypothesis. He was convinced that working in the direction set by those two scholars of metapsychical research would, in a short time, obtain "the right to citizenship in the republic of science" (Preface, ii). The important thing was to avoid at all costs the "maximum danger for the scholar," i.e. "the spiritualism as affective interpretation of some phenomena" (Preface, iii).

The influence of Richet (1922) is evident not only in the choice of the phrase *metapsychical research* instead of *psychical research*, but also in the first four chapters, which are a sort of introduction, free as much as possible from doctrinal comments, to the facts, phenomena, and categories of the discipline. Those chapters follow the path taken by Mackenzie to approach metapsychical research: the thinking animals and Stasia's mathematical mediumship and metapsychical research with its mental and physical phenomena. The fifth and sixth chapters break with the previous method, giving free rein to the philosophical speculations of Mackenzie, who, however, claimed to be aware "of the point at which ends the finding of fact, and begins the commentary around it" (Preface, vi).

In the first chapter Mackenzie returns to the discussion of the thinking animals, particularly the new case of the dog Lola, daughter of the Mannheim's dog. According to his source, Kindermann (1922),[1] Lola was able to understand human conversations, count, speak, and write using proper signs, do weather forecasting, and finally perform small philosophical and moral speeches. Mackenzie felt sure that animals possessed a sort of rudimentary mathematical intelligence anchored in their organic substrate and activated by training; but he was also sure this fact could not explain the philosophical speculations of Lola. Discarding the possibility of fraud, Mackenzie made a step forward compared to his surveys in 1914 and advanced the zoopsychological hypothesis of the "psychic concomitant automatism" (*automatismo psichico concomitante*): The mind of thinking animals, refined by training, would become the receptacle of subconscious thoughts of their masters, or of other people, thanks to the establishment of an automatic and unconscious psychic relationship. The source of all manifestations of intelligence that were not mathematical were attributable to humans (p. 63ff).

Supernormal mathematical skills connect first to the second chapter. In 1913 in a mediumistic circle in Brussels there began to manifest an alleged mediumistic personality who called herself Stasia; initially she proved to be able to influence at will the output of a playing card from a deck, then she began to predetermine the extraction of a specific card, unknown to the investigators until the last moment, through complex arithmetic operations whose results, combined with a code, restored the name of the card that was actually taken from the deck. Thanks to the accounts of Poutet (1919) (p. 67ff), the organizer of the circle in Brussels, and to his séances in 1921 (104–125), Mackenzie guaranteed the authenticity of the mathematical skills of Stasia, whom he recognized as an autonomous subconscious personality, and excluded the skills that were attributable to the conscious intellect of the medium. According to Mackenzie, Stasia also operated through clairvoyance and telekinesis, but those two capacities constituted "so to speak only the skeleton of the events. Around this skeleton, the living pulp is formed by the unprecedented power of calculation Stasia reveals" (77–78).[2]

In Chapter 3 Mackenzie said that after numerous studies with the medium Eusapia Palladino (1854–1918) (e.g., Alvarado 1993, Biondi 1988), who gave life to facts established by a large number of people, the time had come for a decisive step toward the scientific and experimental investigation of mediumistic facts. This meant that metapsychical research after Palladino had to eliminate, as we have already said, any theoretical and emotional contamination with spiritualism, "the vice of origin which weighed heavily on the young branch of study" (p. 148), and embrace the research program conceived by Richet (1922). For Mackenzie, metapsychical research studied the whole phenomenology of *mediumship* or the *supernormal*; the two terms were understood as synonyms but without spiritualist implications (p. 154). The term *supernormal* designated all the activity occurring outside the neuromuscular mechanism of the subject, i.e. veridical perceptions created by mediums without the use of their sense organs, and the actions performed by the mediums, which can be determined objectively, without the use of their body. Recovering the distinction of Richet (1922) between subjective and objective phenomena and adapting over it his own distinction, Mackenzie divided mediumship into "perceptual or static" (in broad terms mental mediumship) and "physical or dynamic" (physical mediumship). In turn these two classes are part of two other higher categories, that of "supermediumistic phenomena" (*fenomeni supermedianici*) and of "mediumistic and submediumistic phenomena" (*fenomeni medianici e submedianici*), based on the distance of the phenomenon from the medium. For example, a phenomenon that takes place away from the medium, such as the clairvoyance of an event in another city, will be a supermediumistic phenomenon, a closer one will be mediumistic (p. 239ff).

Referring to a large and thriving tradition of research on dissociation, multiple personalities, and potentialities of the subconscious mind (e.g., Alvarado 1991, Cardeña and Alvarado 2011), Mackenzie claimed that under certain circumstances during the séances, it is as if the mind of the medium dissociates into at least two parts, the cerebro–spinal psychism, which continues to preside over the normal physiological functions of the individual, and the supernormal psychism, which is totally subconscious and has the power and knowledge to now completely refute any conventional scientific explanation as to supernormal perceptions and actions. Mackenzie wanted to specify that mediumistic dissociation is not entirely comparable to that of hysteria; rather, between the two is a relationship of homology that could reveal a common phylogenetic origin and a natural spread of dissociative phenomena. Therefore, the dissociative forms of mediumship are probably, in embryo, in most normal subjects, both in waking, better still in sleep (dream); they turn out to be more precise in insights, especially when they are very brilliant; can assume greater importance in some states of hypnosis and somnambulism; and, finally, may reach maximum size during the deep trance of the fully developed medium (pp. 167–168).

The third chapter closes with a succinct review of the perceptual mediumship phenomena: telepathy, clairvoyance, psychometry, psychic photography, mediumistic tiptology, automatic writing,[3] polyglot mediumistic phenomena, cross-correspondences, direct writing, and mediumistic dictation. Mackenzie drew from a vast range of sources, thus demonstrating his knowledge of the literature (e.g., Bozzano 1921, Flournoy 1900, Gurney, Myers, & Podmore 1886, Myers 1903, Ochorowicz 1889, Richet 1922, Sage 1904, Warcollier 1921). While he considered cross-correspondences as one of

Title page of
Metapsichica Moderna

the few pieces of evidence strongly in favor of the spiritistic hypothesis, he considered direct writing and mediumistic dictation as phenomena in which counterfeiting and fraud were plentiful. It is worthy of mention, however, that the invitation that he turned to the experimenters do not blame and do not detract too much the mediums found to cheat, because fraud could also be attributed to an expression of their unconscious wishes and aspirations (p. 180ff).

The fourth chapter is devoted to physical mediumship, whose phenomena were for Mackenzie the most numerous and well-documented: "It is certainly an excellent title, especially about the many who tend to deny a priori, [it] can show diagrams of recorders, weight charts, impressions, casts, photographs" (p. 197). Through discussion of some works on ectoplasmy, telekinesis, and, to a lesser extent, on other physical phenomena such as materializations and dematerializations (e.g., those of Bisson (1921),[4] Bozzano (1919), Crawford (1918), Ochorowicz (1889), Schrenck-Notzing (1920), and Zöllner (1878)), Mackenzie found three indisputable experimental results: the existence of a substance that comes out from the medium and forms the ectoplasms, the fact that the output of this substance causes a decrease in the weight of the medium proportional to it, and the intervention of the substance in the phenomena of telekinesis (p. 198ff).

As Alvarado has shown (2006), at the time Mackenzie was writing it was still widely thought that there were forces or substances that go out of the body to produce the phenomena of mental and physical mediumship, forces that seemed to Mackenzie "a new physics that violates, in any way, and from the foundations, all our regular physics, and violates all our fundamental physiological laws" (p. 199). For the author, the central point was the need for someone to explain the way in which the body of the medium disintegrated partially and then recomposed as a new substance, because it was precisely that aspect of the phenomenon that allowed him to present a dissociative/aggregative vision of the mediumship, therefore composed by psychic dissociation and, in the dynamic phenomena, also by physical dissociation.

The fifth chapter and sixth chapters constitute the theoretical (and more original) part of *MM*. Mackenzie is dissatisfied with Geley's (1905) explanation for ectoplasms, due to the somewhat outdated hypothesis of Entelechy described by the vitalist philosopher Driesch (1909). Driesch played a leading role in the revival of Vitalism, which lasted until the 1930s (McLaughlin 2008) and influenced Mackenzie and many other researchers. For Mackenzie the real problem regarding physical mediumship "must be just that of the twofold psychophysical 'dematerialization' of the subject: homologous system, probably, of the twofold dematerialization which takes place in radioactive phenomena of inorganic matter" (pp. 270–271). Remaining doubtful about the nature of the substance and because physical dissociation is dependent on trance mediumship and therefore on psychic dissociation, Mackenzie went on to present his original interpretation of mental mediumship, the naturalistic and anti-spiritical "polypsychical hypothesis" (*ipotesi polipsichica*).

In developing that hypothesis he claimed to have been inspired by Schopenhauer (1925) for the philosophical aspect and by Durand de Gros (1894) for the biological aspect, integrating them with his own studies in those disciplines. Mackenzie argued that every natural being is polyzoic (*polizoico*), or colonial, that is composed of several independent living units that join together and form a new and more articulated being. The polyzoism (*polizoismo*) is naturally occurring from unicellular algae that combine to pave the way for a multicellular organism, to man. The polyzoism brings with it the polypsychism (*polipsichismo*): When the new body is formed from individual minds or psychisms, a new psychic entity emerges that is more complex and of a higher order, greater than the sum of the individual parts and supernormal compared with them. Consider, for example, the apparently intelligent behavior of social insects (p. 276ff). Mackenzie extended the idea that the supernormal can be biological to mediumship: Personalities that manifest themselves during the séances would not be spirits of the deceased, nor even the simple subconscious of the medium or of the individual investigators, but instead the polypsychic entity that emerges from the psychisms of the medium and the sitters, with all the hallmarks of complexity and supernormality that we find in the biological kingdom. In fact, a medium's psychic *quid*, somehow free and active thanks to trance and dissociation, would be able to act on the psychic entities of the experimenters and aggregate them in a completely new polypsychic personality. Once formed, that personality would also produce the paranormal phenomena:

> Such a completely new "mediumistic personality" would therefore in genu-
> ine cases be observed in the best séances and "would animate" the most
> different phenomena: from the spiritual manifestations of the medium to
> the real "ghosts"—not excluding other physical manifestations—clearly re-
> vealing an "intelligence" that models and directs them. (pp. 286–287)

Mackenzie realized that in the explanation of cross-correspondences, identification of mediumistic personality, and haunting phenomena his hypothesis was weaker than that of the spiritists. To solve this problem, he proposed the probable existence of psychic entities unrelated to those present, but still far removed from a deceased person with his own identity and memory (p. 296ff).

The sixth and last chapter is a heterogeneous set of considerations on spiritualism, metapsychical research, philosophy, and science. For example, Mackenzie rebuked spiritualism as being materialistic (p. 309ff), and used the latest discoveries in physics showing increasingly subtle energies and matter to show the robust dualism of the past (p. 315ff). The real dualism, Mackenzie continued, is not between spirit and matter but between the life domain of biology and the non-life domain of physics. "Metapsychical research . . . would be, or at least should aspire to be, the nascent chapter of a widely understood Biology" (p. 338). Moreover, Mackenzie had a psychoanalytical reading of the visions of some mediums, and expressed the idea

> that, perhaps, *psychoanalysis and metapsychical research are two closely re-*
> *lated sisters;* the two youngest legitimate daughters of the great concept of
> the "subconscious," which, in many ways, enlivens the psychological doc-
> trines of man. (p. 343)

MM concludes with indications for the future development of metapsychical research. According to Mackenzie, it should incorporate the psychoanalysis of mediumistic dreams, investigating the desires of the polypsychic supernormal community which are expressed in them; the physical survey, to understand how chemical and physical ectoplasm could exist; and, finally, the biological survey to study the psychic force of the human body (p. 347ff).

Conclusions

The theorist in Mackenzie prevailed over the experimenter, and tried to offer an explanation of paranormal phenomena in a biological and anti-spiritical key. In his view, the spiritistic hypothesis could not definitively rule out the intervention of the living in the production of mediumistic phenomena and could not bring any crucial evidence for spiritism. So he proposed a viable alternative, rooted in biological fact and free of any emotional and religious creeds, to the survival hypothesis.

The mixture of metapsychical research and the philosophy of biology is what is intriguing in Mackenzie's book. We can agree with Talamonti (1971) that Mackenzie's speculations consist of the following points:

1) the priority of the psyche, understood as an element that gives order to biological matter;
2) the existence of a collective psyche for each animal species, which operates on a subconscious level;
3) the existence of psychic fields, from which individuals draw subconscious patterns of behavior;
4) individuals aggregate to form a new polyzoic and polypsychic entity, from which emerge new and more complex behaviors;
5) some polypsychic entities produce mediumistic phenomena.

As we said, for Mackenzie the new polypsychic entity is supernormal compared with the individuals who generate it: For example, the behavior of a colony of termites is supernormal compared to that of the single termite, and is driven unconsciously by the psychic field. In this way, the mediumistic supernormal is only a further degree of that organic supernormal, with all those complex, directed, and non-mechanical behaviors of living beings that occur regularly in nature and are equally difficult to explain. There is no difference between the two. Consequently, mediumship is a biological fact present at all levels of life, without exceptions.

Although *MM* received positive reviews, the polypsychic hypothesis generated some concerns in the research community. For example, in France, Sudre (1923) agreed with Mackenzie's criticisms of spiritualism, but stated that the constancy and coherence of the phenomena, regardless of the composition of the mediumistic circle, denied or made unlikely the polypsychic hypothesis as an explanation for any kind of mediumistic phenomena. In Italy, Bozzano (1923a), moving from a position diametrically opposed to that of Sudre, was in fact a strong supporter of the spiritistic hypothesis, but also had the same criticisms of the polypsychic hypothesis. While recognizing the validity of Mackenzie, of whom he was a friend, he said that according to his case studies he could provide numerous examples of mediumistic personalities who maintained their own and very specific identity during several sessions with different participants, or even in cases of a single participant, thus demonstrating their complete independence from the psychism of the experimenters. With these arguments, Bozzano

(1923b) also wrote to support the spiritical and not polypsychical nature of the mediumistic personality Stasia.

Notes

[1] In the last part of Kindermann (1922) we find an essay by Mackenzie dedicated to the evolution of the thinking animal studies from 1914 to 1919 (Mackenzie 1922), the translation of a booklet originally published in Italian (Mackenzie 1920), in which he argued the same ideas presented in the first chapter of *MM.*

[2] In the Appendix of *MM* there is *Contributo allo studio dell'attività intellettuale subconscia di "Stasia"* (*Contribution to the Study of Stasia's Subconscious Intellectual Activity*) written by the Italian mathematician and logician Alessandro Padoa (1868–1937). He tried to reconstruct, on the basis of reports by Poutet (1919) and Mackenzie (1923), some of the calculations and mathematical reasoning made by Stasia to determine the output of the cards. His conclusions were that the calculations were not prepared and stored by the medium before the séances and that the phenomena, therefore, could be entirely due to Stasia, a real subconscious personality with its own memory and intentionality (Mackenzie 1923: Appendix A, 369–391).

[3] In the Appendix of *MM* we find *Note introspettive del Dott. N. N. sulle proprie sensazioni durante le sue prime prove medianiche* (*Dr. N. N.'s introspective notes on his feelings during his first mediumistic experiences*), an anonymous report of "a prominent scholar" (p. 185) of impressions and physical sensations experienced during some séances in which he developed automatic writing (Mackenzie 1923: Appendix C, 405–411).

[4] In the Appendix of *MM* we find *Relazione della signora J. A. Bisson al 1° Congresso Metapsichico Internazionale (Kopenhagen, 1921)* (*Report of Mrs. J. A. Bisson at the 1st International Metapsychical Conference (Copenhagen, 1921))*. Juliette Bisson was a French sculptor and psychical researcher who studied the mediumship of Eva Carrière for years; in this report, included in the original language, she described Eva's materializations and the substance that produced them. She argued that the phenomena were genuine, that the substance was not ingested and then regurgitated, and that the driving force behind the materializations was intelligent (Mackenzie 1923: Appendix B, 392–404).

LUCA GASPERINI
gasperiniluca@ymail.com

References

Alvarado, C. S. (1991). Iatrogenesis and dissociation: A historical note. *Dissociation, 4,* 36–38.

Alvarado, C. S. (1993). Gifted subjects' contributions to psychical research: The case of Eusapia Palladino. *Journal of the Society for Psychical Research, 59,* 269–292.

Alvarado, C. S. (2006). Neglected near-death phenomena. *Journal of Near-Death Studies, 24,* 131–151.

Beverini, S. (1988). *Introduzione Biografica.* In W. Mackenzie, *Guida ai Fenomeni Medianici (Metapsichica Moderna),* Genoa: Fratelli Melita, 1988, pp. 1–6.

Biondi, M. (1988). *Tavoli e Medium. Storia dello Spiritismo in Italia.* Rome: Gremese.

Biondi, M., & Tressoldi, P. E. (2007). *Parapsicologia: Storia, Ricerca, Evidenze.* Bologna: Il Mulino.

Bisson, J. A. (1921). *Les Phénomènes Dits de Matérialisation.* Paris: Félix Alcan.

Bozzano, E. (1919). *Dei Fenomeni d'Infestazione.* Rome: Luce e Ombra Editions.

Bozzano, E. (1921). *Gli Enigmi della Psicometria.* Rome: Luce e Ombra Editions.

Bozzano, E. (1923a). Considerazioni sull'opera *Metapsichica Moderna* del Dott. William Mackenzie. *Luce e Ombra, 23,* 104–112.

Bozzano, E. (1923b). La storia di Stasia. *Luce e Ombra, 23,* 171–179.

Cardeña, E., & Alvarado, C. S. (2011). Altered Consciousness from the Age of Enlightenment through the Mid-Twentieth Century. In edited by Etzel Cardeña and Michael Winkelman, Editors, *Altering Consciousness: Multidisciplinary Perspectives, Vol. 1: History, Culture, and the Humanities,* Santa Barbara: Praeger, pp. 89–112.

Crawford, W. J. (1918). *The Reality of Psychic Phenomena: Raps, Levitations, etc.* London: J. M. Watkins.

Driesch, H. (1909). *Philosophie des Organischen.* Leipzig: W. Engelmann.

Durand de Gros, J. P. (1894). *Le Merveilleux Scientifique.* Paris: Félix Alcan.

Flournoy, T. (1900). *Des Indes à la Planète Mars: Étude sur un Cas de Somnambulisme avec Glossolalie.* Paris: Le Seuil.

Geley, G. (1905). *L'Etre Subconscient: Essai de Synthèse Explicative des Phénomènes Obscurs de Psychologie Normale et Anormale.* Paris: Félix Alcan.

Gurney, E., Myers, F. W. H., & Podmore, F. (1886). *Phantasms of the Living,* two volumes. London: Rooms of the Society for Psychical Research–Trübner & Company.

Kindermann, H. (1922). *Lola or The Thought and Speech of Animals.* London: Methuen & Co.

Mackenzie, W. (1912a). *Alle Fonti della Vita. Prolegomeni di Scienza e d'Arte per una Filosofia della Natura.* Genoa: A. F. Formiggini Editions.

Mackenzie, W. (1912b). I cavalli pensanti di Elberfeld. *Rivista di Psicologia applicata alla Pedagogia ed alla Psicopatologia, 8,* 479–517.

Mackenzie, W. (1914). *Nuove Rivelazioni della Psiche Animale (da Esperimenti dell'Autore).* Genoa: A. F. Formiggini Editions.

Mackenzie, W. (1915). *Significato Bio-Filosofico della Guerra.* Genoa: A. F. Formiggini Editions.

Mackenzie, W. (1920). *Gli Animali "Pensanti" e l'Ipotesi dell'Automatismo Concomitante.* Genoa: G. B. Marsano Print Shop.

Mackenzie, W. (1922). *"Thinking" Animals: A Critical Discussion of Developments from 1914 to 1919.* In H. Kindermann, *Lola or The Thought and Speech of Animals,* London: Methuen & Co., pp. 157–188.

Mackenzie, W. (1923). *Metapsichica Moderna: Fenomeni Medianici e Problemi del Subcosciente.* Rome: Libreria di Scienze e Lettere.

Mackenzie, W. (1967). *Les Grandes Aventures Spirituelles.* Paris: Encyclopédie Planète.

Mackenzie, W. (1988). *Guida ai Fenomeni Medianici (Metapsichica Moderna).* Genoa: Fratelli Melita. http://spiritismo.altervista.org/classici.htm

Marhaba, S. (2003). *Lineamenti della Psicologia Italiana: 1870–1945.* Florence: Giunti.

McLaughlin, B. (2008). Vitalism and Emergence. In T. Baldwin, Editor, *The Cambridge History of Philosophy 1870–1945*, Cambridge: Cambridge University Press, pp. 631–639.

Morselli, E. (1908). *Psicologia e Spiritism: Impressioni e Note Critiche sui Fenomeni Medianici di Eusapia Paladino* (two volumes). Turin: Fratelli Bocca.

Myers, F. W. H. (1903). *Human Personality and Its Survival of Bodily Death* (two volumes). London: Longmans Green & Co.

Ochorowicz, J. (1889). *De la Suggestion Mentale*. Paris: Octave Doin.

Poutet, H. (1919). Phénomènes psyco-physiologiques à Bruxelles. *Annales des Sciences Psychiques, 19,* 54–58.

Richet, C. (1922). *Traité de Métapsychique*. Paris: Félix Alcan.

Sage, M. (1904). *Mrs. Piper & the Society for Psychical Research*. New York: Scott-Thaw Co.

Schopenhauer, A. (1925). Magnetismo Animale e Magia. In *Memoria Sulle Scienze Occulte*, pp. 1–48, Milan: Fratelli Bocca. [Originally published as *Über den Willen in der Natur* in 1836]

Schrenck-Notzing, A. (1920). *Physikalische Phaenomene des Mediumismus: Studien zur Erforschung der Telekinetischen Vorgänge*. Monaco: Ernst Reinhardt.

Sudre, R. (1923). Review of Mackenzie, William. *Metapsichica Moderna: Fenomeni Medianici e Problemi del Subcosciente*. Rome: Libreria di Scienze e Lettere. *Revue Métapsychique, 3,* 193.

Talamonti, L. (1971). Ricordo di William Mackenzie (1877–1971). *Rassegna Italiana di Ricerca Psichica, 2–3,* 7–18.

Warcollier, R. (1921). *La Télépathie: Recherches Expérimentales*. Paris: Félix Alcan.

Zocchi, P. (no date). I Cavalli Pensanti di Elberfeld. *Archivi Storici della Psicologia Italiana*. http://www.archiviapsychologica.org/index.php?id=611

Zöllner, J. K. F. (1878). *Wissenschaftliche Abhandlungen*. Leipzig: Staackmann.

Journal of Scientific Exploration, Vol. 26, No. 4, pp. 923–925, 2012   0892-3310/12

Prescribing Sunshine: Why Vitamin D Should Be Flying Off Shelves
by M. Aziz. CreateSpace, 2012. 250 pp. $7.99 (paperback). ISBN 978-1478396079. Kindle Edition $2.99.

Aziz was led by circumstances in his own family to realize the neglected importance of vitamin D. Although this account of his experience is written in a very personal style, most people will benefit from several aspects of the book. A general one is the fact that the medical profession is not always sufficiently knowledgeable and helpful, in part because different practitioners take different views and because some of those views are not necessarily well-informed from the primary literature of medical science. A more specific benefit of this book comes from the cornucopia of citations to the scientific literature which reveal things about vitamin D that I imagine few people are aware of.

The general point, that advice from the medical profession may be doubtfully informed by the medical-science literature, is particularly pertinent when it comes to diet. Aziz makes the same point with respect to vitamin D as Linus Pauling made about vitamin C and vitamins more generally: Medical science knows only how much is needed to prevent actual manifest ill-health, and this may be much less than the optimal amount required for optimal health. I'm reminded of the experience of Colin Campbell, author of *The China Study*, whose blurbs include "The most comprehensive study of nutrition ever conducted—startling implications for diet, weight loss, and long-term health": Campbell once told me that all his research had been funded through grants to study cancer because there is so little in the way of research funds for direct studies of nutrition.

Some of what Aziz learned about vitamin D connects in one way or another with heart disease and cholesterol, and with HIV/AIDS, and with auto-immunity and with flu and with cancer. Many of the inferences from cited facts are speculative, as the author acknowledges quite explicitly; nevertheless, the cited sources allow readers to re-examine and take these notions further. For instance, Aziz notes that longer telomeres (appendages on chromosomes) have been associated with longer lifespan; and cites a report that high levels of vitamin D are associated with telomeres not shortening—could vitamin D be an elixir of longer life? Follow the citation and read and speculate further for yourself.

One of Aziz's citations leads to the results of a 10-year study by the World Health Organization (WHO) of cholesterol and heart disease, which found no correlation between them (Kendrick 2007). Many others have deconstructed the myth that cholesterol *causes* heart disease (Kauffman 2006), but the WHO data seem finally conclusive on this point. A number of sources have discussed the damaging "side" effects of statins, but few have pointed out that cholesterol plays an important part in health, for example because it is an essential component of cell walls. Too little may cause harm; indeed, it has even been suggested that artificially limiting cholesterol could be a cause of Alzheimer's disease (Lorin 2005). Moreover, when substances are made in the body, there are feedback mechanisms to regulate proper rates and amounts. The optimums may differ from individual to individual, so it seems unwarranted to declare someone's cholesterol (or anything else) as too high (or too low, or too LDL or HDL, etc.) just because it differs from some population average.

One of the things I learned from this book is that, chemically speaking, vitamin D is more complex a molecule and more physiologically powerful than the other vitamins, which are much simpler chemicals. Vitamin D is a steroid and thereby akin to hormones and pervasively active physiologically. Further, though I had known that we make our own vitamin D through exposure to sunshine, I had not known that the precursor from which the D is made is cholesterol. Therefore, Aziz points out, artificially limiting cholesterol might also contribute to deficiency of vitamin D.

As to the connection between vitamin D and (resistance to) flu: Why is the incidence of influenza seasonal, if that is not connected to the degree to which sunlight produces vitamin D for us?

So there are many intriguing questions opened up by Aziz's citations and speculations. His personal story also illustrates the enormous difficulty of trying to test things on oneself; yet given the unreliability of much official advice, we may well feel the need to do so and to trust to our own instincts about what the primary literature says. After all, the unreliability of official advice has been noted even by a former editor of the *New England Journal of Medicine* (Angell 2009):

> It is simply no longer possible to believe much of the clinical research that is published, or to rely on the judgment of trusted physicians or authoritative medical guidelines. I take no pleasure in this conclusion, which I reached slowly and reluctantly over my two decades as an editor of *The New England Journal of Medicine.*

A pervasive error in much of the medical literature is the confusion of correlations, associations, with causation. "Risk factors" are associations, symptoms, but mainstream practice persistently takes them to be causes, whereupon drugs are prescribed to lower blood pressure and cholesterol and blood sugar and PSA measures just because those things are associated with morbidity. But those things are all associated also with age, and correlated with one another in the first instance because of that. To prove causation would require data that are not yet in hand, for treatments based on the belief in causation have not been shown to be beneficial: "There are no valid data on the effectiveness" of "statins, antihypertensives, and bisphosphanates" (the last are prescribed against osteoporosis; Järvinen et al. 2011).

Almost all that's available about human health consists of associations, given that we shy away from conducting actual experiments on human beings. So Aziz too speculates on the basis of associations, and like the medical establishment and everyone else must use judgment as to the plausibility that any given association is more than a mere *coincidence*, a non-causal correlation. Readers will not always agree with the judgments suggested in this book, but they, and particularly their sources, are well worth attending to.

HENRY H. BAUER
Professor Emeritus of Chemistry & Science Studies
Dean Emeritus of Arts & Sciences
Virginia Polytechnic Institute & State University
hhbauer@vt.edu; www.henryhbauer.homestead.com

References

Angell, M. (2009). Drug Companies and Doctors: A Story of Corruption. *New York Review of Books, 56*(1), 15 January.

Järvinen, T. L. N., Sievänen, H., Kannus, P., Jokihaara, J., & Khan, K. M. (2011). The true cost of pharmacological disease prevention. *British Medical Journal, 342*(d2175), 1006–1008. doi 10.1136/bmj.d2175

Kauffman, J. M. (2006). *Malignant Medical Myths: Why Medical Treatment Causes 200,000 Deaths in the USA Each Year, and How to Protect Yourself.* West Conshohocken, PA: Infinity. pp. 78–104 (Myth 3).

Kendrick, M. (2007). The Cholesterol Myth Exposed. http://www.youtube.com/watch?v=i8SSCNaaDcE

Lorin, H. (2005). *Alzheimer's Solved.* BookSurge/Create Space. http://www.booksurge.com

Journal of Scientific Exploration, Vol. 26, No. 4, pp. 926–927, 2012   0892-3310/12

Unexplained! Strange Sightings, Incredible Occurrences, and Puzzling Physical Phenomena by Jerome Clark. Canton, MI: Visible Ink Press, 2012 (third edition). 600 pp. $22.95 (paperback). ISBN 978-1578593446; Kindle 978-157859290; ePub 978-1578594283; PDF 978-1578594276.

Clark is widely known and highly respected for authoritatively reliable and judicious discussion of anomalous phenomena. This volume is a treat for connoisseurs of the genre, and should be read by everyone interested in trying to understand unexplained and often seemingly *unexplainable* matters.

The Introduction of 11 pages is worth reading and re-reading. It reflects Clark's considered judgments, which I don't hesitate to describe as wisdom

acquired over decades of grappling with reports and other evidence about matters that most of mainstream science and scholarship find too difficult to handle.

Clark begins by acknowledging that human beings find it very difficult to say, "I don't know." Yet that is the proper conclusion about many unexplained matters. Often it is the most expert and the most knowledgeable specialists who have the confidence to say it, and Clark exemplifies that.

Although this is labeled a third edition, it really is a new book, as the Foreword explains. There are fewer than half as many individual topics as in the first edition, and they are treated discursively in essay fashion rather than, as in earlier editions, as shorter encyclopedia-type pieces. The present approach allows Clark to underscore with particular cases the general insights offered in the Introduction.

The contents are organized into three parts: Mysteries, Curiosities, Fables.

Mysteries include Anomalous Clouds; Black Dogs; Lake Monsters; Living Dinosaurs; a couple of dozen topics in all. The evidence for the reality of reports is generally considerable, and reasonably possible explanations are often suggested. Curiosities include some that have been entirely debunked as spurious, like Green Children or the Jersey Devil, but also a cryptozoological creature, the Onza, that turned out to be quite real albeit "only" a distinct variant of a known species. Among Fables one finds largely claims that never really deserved the credence some gave them, like the Cottingley Fairy photographs or the assertion of Hollow Earth.

This book deserves to be on one's nightstand and subsequently on one's shelves. The many topical essays offer innumerable points of interest, and the Introduction bears re-reading at intervals. Readers already familiar with Clark's work can reliably expect to relish this volume, and anyone not yet familiar with Clark would do well to begin their acquaintance with this book.

HENRY H. BAUER
Professor Emeritus of Chemistry & Science Studies, Dean Emeritus of Arts & Sciences
Virginia Polytechnic Institute and State University
hhbauer@vt.edu, www.henryhbauer.homestead.com

Journal of Scientific Exploration, Vol. 26, No. 4, pp. 928–930, 2012   0892-3310/12

The Number Sense: How the Mind Creates Mathematics by Stanislas Debaene. New York: Oxford University Press, 1997. 288 pp. $37.95. ISBN 978-0195132403.

What Stanislas Debaene dubs "the number sense" is a natural ability humans share with other animals, enabling us to "count" to four virtually instantaneously. This so-called "accumulator" provides "a direct intuition of what numbers mean" (p. 5). Beyond four, our ability to *perceive* numbers becomes approximate, though concepts enable us to move beyond approximation. Because humans typically learn number concepts in early childhood, we easily forget that our brains retain the number sense throughout life. This book examines the biological basis for this intuitive ability, with nine chapters organized into three readily graspable groups of three. Aside from its frustrating lack of a clear referencing system (making it difficult or impossible to trace Debaene's sources), the book is a pleasure to read.

Part I examines our Numerical Heritage, focusing on animals, human babies, and adult humans, respectively. Chapter 1 recounts stories of various gifted animals whose actions suggested amazing aptitude with numbers. Debaene remains duly skeptical, noting that animals easily draw cues from human trainers informing them (perhaps inadvertently) how to answer. However, evidence from recent scientific experiments demonstrates that (for example) animals know the difference between the way numbers "add up" and the way shapes and colors operate. Some even seem to perform "an internal computation not unlike the addition of two fractions" (p. 25)! Debaene sometimes writes somewhat loosely, giving the impression that animals actually *count*, as if they possess number *concepts*. But only some primates (e.g., chimpanzees) possess such abilities—and only in laboratories, never in the wild (p. 39). His main aim, therefore, is to provide a theory of "how it is possible to count without words" (p. 28). He theorizes that the brain's neural network represents numbers in a "fuzzy" way, enabling animal brains to approximate, whenever the given stimuli require comparison of quantities greater than four.

Debaene could have improved his argument, in Chapter 1 and throughout the book, by distinguishing more explicitly between the function of what Kant calls *concepts* and *intuitions*. The number sense involves only the latter; yet Debaene sometimes inadvertently blurs this distinction—e.g., by using the word *irrational* (p. 27) where *counterintuitive* would be more accurate. A good example of this minor weakness comes in Chapter 2: After surveying Piaget's groundbreaking work on infant numerosity, Debaene claims that "Piagetian tests cannot measure children's true numerical competence" (p. 45), but does not clearly explain why. Expressed in Kantian terms, such tests measure *concepts*, not intuitions. By not making this distinction explicit, and continuing to use "concept of number" (e.g., p. 47) as if it also describes what his own impressive laboratory tests measure in children (whose

details are also explained in Chapter 2), Debaene obscures the fact that his experiments *tested the intuitions* (i.e. the number *sense*) of children, including very young babies. Piaget tested children's *conceptions*; Debaene tested children's *perceptions*.

Chapter 3 examines historical evidence for the number sense in various ancient counting systems, then describes experiments on adults that demonstrate a clear increase in the length of time required to compare pairs of small quantities, relative to the length required to compare pairs of larger quantities. Debaene's rather complex argument at this point could be expressed more simply, using the above-mentioned Kantian distinction: The adult human's brain retains and continues to employ an *intuitive grasp of number* (i.e. the number *sense!*) even after the person has fully developed a working understanding of the *concept* of number. What Debaene's experiments demonstrate, then, is that intuition works faster than conception. That his focus is on intuition, not concepts, becomes even clearer when Chapter 3 concludes with some intriguing speculations on whether numbers might possess not only spatial characteristics (cf. the "number line"), but also colors, shapes, and even sounds. Readers might remain unconvinced unless they interpret such claims as relating to non-conceptual *intuitions*.

Part II moves beyond the number sense and focuses on issues relating more properly to the concept of number as such. Chapter 4 offers a wealth of information about the language of numbers, as manifested in a wide range of cultures. Insights abound: For example, Chinese students are better at mathematics because they can count to ten using only ten syllables; and a *centi*pede has only 42 legs (100 being a conceptual approximation of the perception). Read the text itself to appreciate its rich fare! Chapter 5 complements Chapter 2 with further reflections on child numerosity, focusing this time much more on number *concepts* than on number *intuitions* (i.e. the number sense as such). Here Debaene's text again suffers from a tendency to conflate these factors, as when he refers to children exhibiting "an intuitive understanding of what calculations mean" (p. 124). (Kantians, dry your tears!) Chapter 5 concludes by encouraging teachers to use intuition more effectively in teaching mathematics. Chapter 6, Geniuses and Prodigies, considers whether such cases arise more from special intuitive abilities, or greater conceptual rigor. In answering his (oft-repeated) question, "Where does mathematical talent come from?" (e.g., p. 170), Debaene considers a wide range of intriguing possibilities, regularly (and admirably) admitting his ignorance of any ultimate answer.

Part III, Of Neurons and Numbers, reads almost like three (interesting!) appendices. Chapter 7 reviews a range of cases where brain damage affected a person's number sense in various ways. Emphasizing the brain's plasticity, Debaene repeatedly notes that brain science still leaves much unknown territory. Chapter 8 explores numerous facts about how we *use* the brain in calculating, the brain being "responsible for almost a quarter of the energy expended by the entire body" (p. 211). Chapter 9 concludes the book with some philosophical musings on what numbers *are*. Assuming that "the human brain . . . give[s] birth to mathematics" (p. 232), Debaene insists on abandoning the computer metaphor (p. 237): "The brain is not a logic machine but an analog device." Considering numerous options from various past philosophers, he expresses frustration (p. 240) that "all our attempts to provide a formal definition [of number] go nowhere." Although Debaene might be out of his depth here, a clear intuition–conception distinction would have significantly enhanced his success in attempting to forge "a reconciliation between Platonists and intuitionists" on this issue (p. 251).

STEPHEN PALMQUIST

Department of Religion and Philosophy, Hong Kong Baptist University
Kowloon, Hong Kong SAR, China
stevepq@hkbu.edu.hk

Journal of Scientific Exploration, Vol. 26, No. 4, pp. 931–934, 2012   0892-3310/12

Where Mathematics Comes From: How the Embodied Mind Brings Mathematics into Being by George Lakoff and Rafael E. Núñez. New York: Basic Books, 2000. 512 pp. $29.99. ISBN 978-0465037711.

This book announces a new academic discipline, the "cognitive science of mathematics" (p. xi), by demonstrating how an empirical examination of the *ideas* underlying our use of mathematical symbols and calculations must employ metaphors grounded in the "embodied mind." The authors ruthlessly attack what they dub "The Romance of Mathematics" (p. xv), their metaphor for any approach that treats mathematics as grounded in an abstract, disembodied yet objective reality that mysteriously provides the essential structure to the natural, human world.

The authors declare their a priori assumptions in the Introduction, the most essential being that "human mathematics . . . is an empirical scientific question, not a mathematical or a priori philosophical question" (p. 3). Solely on this basis can they claim that cognitive science, *and cognitive science alone*, answers the question posed by the book's title. Repeatedly referring to this and other central claims as "arguments," the authors actually take as their foundational *presupposition* that "whatever 'fit' there is between mathematics and the world occurs in the minds of scientists" (p. 3). Their rejection of "Platonic mathematics" (p. 4) is obviously circular: The conclusion affirms what the first premise assumed, that "transcendent Platonic mathematics" cannot be "human mathematics." (This tendency toward circularity pervades the book, as when the authors conclude that mathematics, which they have assumed is necessarily grounded in "the cognitive apparatus" [p. 30], turns out to be "not independent" of that apparatus.) The authors never acknowledge that *anything* transcendent, once it is made known to us, must (by definition) make use of metaphors and/or other symbolic processes; so the presence of such metaphors cannot disprove an original transcendence. Following this dubious starting point, an irony colors the entire book: After making such a concerted effort to debunk Platonic (transcendent) mathematics, by demonstrating how all mathematical truths require metaphors in order to be understood by humans, their demonstration could also explain how *Platonic* mathematics comes to be known by us! *This* issue, the question of whether or not there *is* a transcendent realm wherein mathematical truth might be said to "exist," is entirely philosophical; as such, it necessarily transcends this book's empirical, scientific mandate.

The book's 16 chapters are grouped into five parts, with a sixth part reporting a four-part case study. Chapters 1–4, on the Embodiment of Basic Arithmetic, contain little original research, being primarily a summary of some basic tools of "second generation" cognitive science, grafted into reports on the findings of various scientists, regarding the innate abilities of human beings (and some animals) to count up to small numbers (or "subitize") and perform various other simple mathematical functions. My review of one of this book's primary sources, Debaene's *The Number Sense* (also the likely answer to the question: Where did the authors' *title* come from?), is being published together with the present review and covers a similar range of empirical issues, so I shall not comment further on such details here.

The authors sometimes attribute cognitive mechanisms to the brain too hastily (without evidence or argument): e.g., they state that a type of *metaphor*, associating simple arithmetic with "object collection," "arises naturally in our brains" (p. 60; cf. p. 78). But do metaphors themselves really exist *in the brain*? Likewise, they claim arithmetic *arises out of* the "4Gs" (four "grounding metaphors for arithmetic")—The Zero Collection Metaphor (p. 64), The Zero Object Metaphor (p. 67), The Measuring Stick Metaphor (p. 68), and Arithmetic as Motion Along a Path (p. 71)—yet they offer nothing close to an argument that would prove these metaphors are necessary for the very *possibility* of human arithmetic. Provided we remember that the authors' approach is descriptive, the high-sounding conclusions regarding their claims to have shown where arithmetic "comes from" can be accepted for what they are: *empirical* truth-claims. The 4Gs may correctly identify the empirical factors that "make the arithmetic of natural numbers natural" (p. 92) and thus be "constitutive of our fundamental understanding of arithmetic" (p. 94). Yet this does not prove that mathematics has no independent, transcendent status—any more than explaining how one learns the concept "God" would prove that no divine being exists outside of human religious traditions. The authors ignore the distinction that begins Kant's *Critique of Pure Reason*: The fact that "all our knowledge begins with experience" does not imply that "all our knowledge arises out of experience." Conflating "arithmetic" with "our understanding of arithmetic" (e.g., pp. 94–95), they assume that by uncovering the cognitive roots of the latter they are answering the "big question" of the former's source.

The same error pervades Part II, where the authors wade into the deeper mathematical waters of algebra, logic, and set theory. They assume the task of discovering "what algebra *is*, from a cognitive perspective" (p. 119) exhausts the question of where algebra itself *comes from*. Likewise, they reduce Boolean logic and set theory to expressions of mathematical metaphor. They declare mathematical concepts such as "the universal class" to have no "objective existence at all," but to be "created" by an underlying metaphor (p. 131), and insist that "rules of inference" can "preserve truth" only because of such metaphors. Here, as throughout the book, the explanations are often intriguing, yet also frustratingly repetitious.

Whereas most publications in the Lakoff school of cognitive science are filled with an impressive array of metaphors, the empirical observations and scientific conclusions advanced in this book all stem from one foundational metaphor (p. 161), presented in various forms. Only in Part III (The Embodiment of Infinity) does this Basic Metaphor of Infinity (BMI) (pp. 8f.) come to the fore. Chapter 8 unpacks its structure by elucidating "the embodied source of the idea of infinity" in human "*aspectual systems*" (p. 155). A consistent defect throughout their discussion is the tendency to make ontological claims (e.g., "Wherever there is infinite totality, the BMI is in use" [p. 175]), when their argument only justifies *epistemological* claims regarding the necessity of metaphor for *human reference systems*. Chapters 9–11 persuasively argue that our use of real numbers, transfinite numbers, and infinitesimals also relies on the BMI.

Part IV presents a series of historically based accounts of The Discretization Program That Shaped Modern Mathematics (p. 257f.). As intellectual history, their claims about the centrality of metaphor in the various paradigm shifts they analyze may be questionable; but they undoubtedly expose numerous fascinating tendencies, highlighting some insightful implications for the way mathematics ought to be taught.

Part V concludes the main text by discussing "Implications for the Philosophy of Mathematics" (p. 335f.). Here the subtle circularity of the authors' overall argument comes to the fore, as they claim to have dismissed the possibility of any mathematical entities actually existing. In argument after argument they assume that the metaphorical structure of all human thought processes proves that things (e.g., numbers) in themselves are impossible. But what they attack is a straw man: the (silly) claim that numbers somehow exist "out in space" (p. 345).

Taken as an extended exercise in what the authors aptly describe as "mathematical idea analysis" (p. 29 et al.), this book is a tour de force for cognitive science. Despite the authors' reluctance to acknowledge any predecessors, it defends an essentially Kantian thesis: The human mind itself *creates* mathematics. Yet the authors badly err by inferring from their own self-confessed presuppositions that their analysis *proves* mathematics to be *nothing but* "a mere historically contingent social construction" (p. 9). By taking this (unnecessary) extra step, mysteriously insisting—as if it were a magical incantation!—that "[t]here is nothing mysterious, mystical, magical, or transcendent about mathematics" (p. 377), they ironically end up defending an entirely non-empirical thesis that might be called *the Romance of the Cognitive Science of Mathematics*.

Stephen Palmquist
Department of Religion and Philosophy, Hong Kong Baptist University
Kowloon, Hong Kong SAR, China
stevepq@hkbu.edu.hk

Journal of Scientific Exploration, Vol. 26, No. 4, pp. 935, 2012   0892-3310/12

Wari: Lords of the Ancient Andes by Susan E. Bergh et al. New York: Thames & Hudson, 2012. 304 pp. $60 (hardcover). ISBN 978-0500516560.

Virtually everyone has heard of the Inca Empire, centered in Cuzco, Peru, and some are at least vaguely familiar with Bolivia and Peru's earlier Tiwanaku (Tiahuanaco) culture. Few, on the other hand, have heard of Tiwanaku's powerful contemporary, the Wari, centered in Ayacucho, Peru (circa A.D. 500–1300). This important volume is the only major and comprehensive work on these people of which I am aware, and was issued in connection with a Cleveland Museum of Art–organized traveling exhibition (the Museum of Art Fort Lauderdale, the Kimball Art Museum in Fort Worth) of Wari portable material culture. It should correct, to a significant degree, widespread ignorance of the Wari. The culture's history and the minimal history of its investigation receives attention, as do its beliefs, deities (especially staff gods), rituals, feasting, and built-environment (the latter architecturally far less spectacular that that of Tiwanaku or the Inca, perhaps accounting for the Wari's low profile in contemporary awareness). There follow lavishly illustrated chapters on decorated ceramics (including effigy vessels), textiles, featherwork, metal and inlaid objects, figurines, and wooden containers. Some of these esthetics will appeal strongly to Western tastes. Some of the textile designs, although highly disciplined, have an almost modernistic, abstract look.

I find the tie-dyed textiles to be of particular interest, because tie-dyeing, especially that involving lines of dot-in-square motifs, is shared with parts of the Old World, especially in southern Asia, as well as with greater Mesoamerica, and is associated with religious and/or political authority. Tie-dye dates to well before the Christian Era in Peru. Intriguingly, many Wari tie-dyed cloths are constructed in patchwork arrangements. In Java, (non-tie-dyed) patchwork garments were worn by priests and rulers for protection. I have suggested transpacific importation of the method from Southeast Asia to South America, although a reverse direction of diffusion is also conceivable (Jett 1999).

STEPHEN C. JETT

scjett@hotmail.com

Note: This review will also appear in *Pre-Columbiana: A Journal of Long-Distance Contacts*.

Reference

Jett, S. C. (1999). Resist-dyeing: A possible ancient transoceanic transfer. *NEARA Journal, 33*(1), 41–55. Pembroke, NH: New England Antiquities Research Association.

Journal of Scientific Exploration, Vol. 26, No. 4, pp. 936–938, 2012   0892-3310/12

Radical Theory of First Americans Places Stone Age Europeans in Delmarva 20,000 Years Ago by Brian Vastag. Washington, D.C.: *The Washington Post*, February 29, 2012.

New Evidence Suggests Stone Age Hunters from Europe Discovered America. London: *The Independent*, February 28, 2012.

Iberia, Not Siberia? by David Malakoff. *American Archaeology*, *16*(2), 2012, 38–44.

Critics Assail Notion That Europeans Settled Americas by Michael Balter. *Science*, *335*(6074), 2012, 1289–1290.

In 1999, archaeologist Dennis Stanford, of the Smithsonian Institution's National Museum of Natural History, resurrected and elaborated the idea that the ancestors of North America's Late Pleistocene Clovis people were Solutreans of the European Upper Paleolithic. Almost all professionals rejected this notion, which was based on notable similarities between the lithic artifacts of the two manifestations. Now, with Exeter University (UK) archaeologist Bruce Bradley, Stanford has put out a major book displaying the hypothesis and arguing for it (Stanford & Bradley 2012, to be reviewed in a later issue), and a certain number of their colleagues are taking this striking suggestion seriously. Most recently, Stanford points especially to lithics recovered from six mid-Atlantic sites, three of which are in the Eastern Shore area of the Delmarva Peninsula to the east of Chesapeake Bay. The implements look very like European ones dating to between 19,000 and 26,000 years ago.

Previously, the earliest American tools showing such similarities had been no more than 15,000 years old, much later in time than the Solutrean ones they resembled, and most specialists felt that this time gap was unbridgeable. However, data diminishing the time-gap problem have been accumulating. In 1970, a Virginia trawler hauled up a 22,000-year-old mastodon bone and a nearly 20-centimeter-(8-inch-)long stone blade from the bottom of the Atlantic 60 miles offshore—a locale that would have been dry land during the last Ice Age. The date of the bone was nearly twice as great as the age of the Clovis archaeological culture, which was then almost universally believed to be a manifestation of the first humans to have entered the New World, via the then-dry Bering Strait region. More recently, sediments in Maryland dated to as much as 25,000 years ago have yielded projectile points resembling Solutrean ones from Europe of similar age. And a European-style stone knife found in Virginia in 1971 has proven to be made from a chert originating in France.

Stanford and Bradley's proposal—known as the "Iberia, not Siberia" hypothesis—is that during the last glacial maximum (LGM), Solutreans made their way for some 1,500 miles along the food-rich edge of the North Atlantic sea ice from Europe to North America, perhaps in skin boats, feeding on seals, fishes, and sea birds. During the time period indicated, Northeast Siberia was devoid of humans, precluding a Beringian route of entry to the hemisphere. The main ancestors of today's American Indians would have entered millennia later, from northeastern Asia.

From its initial presentation, Stanford and Bradley's hypothesis has drawn harsh criticism in a field historically full of unpleasant contentiousness. For instance, the archaeologist Ted Goebel alleges that the idea was "dead on arrival long before" the book appeared (Balter:1289). In addition to the time gap mentioned above, this attitude is based largely on the fact that genetics clearly shows an overwhelmingly northeastern Asian origin of today's Native Americans as well as for fossil humans in America back to as much as 14,000 years ago. But there are no certain Clovis or clearly pre-Clovis skeletons to test, and critics seem not to consider the possibility that the carriers of Solutrean culture were genetically swamped by later, Asian, arrivals, and contributed little to the genetic makeup of later American Indians or Eskimos. And opponents prefer to explain the North American Indian minority presence of the European mitochondrial-DNA haplogroup X as somehow having been carried by land across Eurasia and Beringia and having later disappeared in Siberia rather than its having been carried across the Atlantic Ocean.

European archaeologists further point out that there is no evidence of a Solutrean maritime lifeway—although any such evidence would have been largely covered by post-Pleistocene rising sea levels. The implication of some climate models that, during the LGM, sea ice was not continuous across the North Atlantic during most of the year seems to me implausible in light of more-recent historic distribution of sea ice. Too, inadvertent westward drift of a group of humans on a detached ice floe is an alternative to intentional migration along the ice margin.

My personal sense at this point is that the Stanford–Bradley hypothesis is quite plausible but meets strong resistance not only for not-entirely-thought-out evidentiary reasons but even more because it differs so dramatically from existing paradigms.

CR. STEPHEN C. JETT
scjett@hotmail.com

Reference

Stanford, D. J., & Bradley, B. A. (2012). *Across Atlantic Ice: The Origins of America's Clovis Culture.* Berkeley: University of California Press.

32nd ANNUAL MEETING OF THE SOCIETY FOR SCIENTIFIC EXPLORATION
DEARBORN (DETROIT), MICHIGAN JUNE 5–JUNE 8 2013
UNSETTLED SCIENCE

Aerial Anomalies (UFOs)—Eddie Bullard; Michael Swords
Historical Anomalies (Pre-Columbian Contacts)—Stephen C. Jett
Mind Anomalies (Levitation, Macro-PK, Yoga)—Donald J. DeGracia
Evening Panel—Do topics under the SSE umbrella have enough in common
    that we can learn from each other?

CALL FOR PAPERS: Send abstracts of 300–500 words by February 15, 2013, to
    Huyghe@anomalist.com

The 2013 Annual SSE meeting will be held at the historic Dearborn Inn, in Dearborn, Michigan, beginning with the Wednesday evening Reception June 5, and running through the Saturday evening Banquet June 8. The Inn is 12 miles from Detroit Metro Airport.

CONFERENCE HOTEL: Dearborn Inn Marriott, 20301 Oakwood Blvd., (313) 271-2700/ SSE rate $110 per night (code SSESSEA). Smoke-free environment, in-room Internet $12.95/day, free parking, gym, outdoor pool, 2 restaurants. The Inn was built in 1931 by Henry Ford. SSE will be in the Alexandria Ballroom.

CUTOFF DATE FOR THE SSE ROOM RATE IS APRIL 27, 2013. SSE rooms will probably fill up long before that.
http://www.marriott.com/hotels/travel/dtwdi-the-dearborn-inn-a-marriott-hotel/
http://www.marriott.com/hotels/travel/dtwdi?groupCode=ssessea&app=resvli
nk&fromDate=6/4/13&toDate=6/9/13

OTHER HOTELS: There are more economical accommodations at the nearby Comfort Inn, 20061 Michigan Ave., Dearborn, which offers complimentary breakfasts and shuttle service to nearby locations, such as local restaurants and the Henry Ford Museum. There are many other nearby hotels.
http://www.comfortinn.com/hotel-dearborn-michigan-MI385.

WELCOME RECEPTION: Wednesday evening June 5th, 6:00–9:00, free
FRIDAY FIELD TRIP: @$60 including lunch

PROGRAM COMMITTEE: Patrick Huyghe (Chair), Erik Schulte, Henry Bauer

LOCAL HOST: Carl Medwedeff SSE2013LocalHost@scientificexploration.org

Society for Scientific Exploration

Executive Committee

Dr. Bill Bengston
SSE President
St. Joseph's College
Patchogue, New York

Professor Robert G. Jahn
SSE Vice-President
School of Engineering and Applied Science
Princeton University
Princeton, New Jersey

Dr. Mark Urban-Lurain
SSE Secretary
College of Natural Science
Michigan State University
111 N. Kedzie Lab
East Lansing, Michigan 48824

Dr. York Dobyns
SSE Treasurer
Lexington, Kentucky

Ms. Brenda Dunne
SSE Education Officer
International Consciousness Research
    Laboratories
Princeton, New Jersey

Professor Garret Moddel
SSE Past President
Department of Electrical and Computer
    Engineering
University of Colorado at Boulder
Boulder, Colorado

Council

Dr. Marsha Adams
International Earthlights Alliance
Sedona, Arizona

Dr. Julie Beischel
The Windbridge Institute
Tucson, Arizona

Dr. Richard Blasband
Center for Functional Research
Sausalito, California

Dr. Roger Nelson
Global Consciousness Project
Princeton University, Princeton, New Jersey

Dr. Dominique Surel
Evergreen, Colorado

Dr. Chantal Toporow
Northrup Grumman
Redondo Beach, California

Appointed Officers

P. David Moncrief
JSE Book Review Editor
Memphis, Tennessee

L. David Leiter
Associate Members' Representative
Willow Grove, Pennsylvania

Erling P. Strand
European Members' Representative
Østfold College, Halden, Norway

Professor Peter A. Sturrock
SSE President Emeritus and Founder
Stanford University, Stanford, California

Professor Charles Tolbert
SSE President Emeritus
University of Virginia, Charlottesville, Virginia

Dr. John Reed
SSE Treasurer Emeritus
World Institute for Scientific Exploration
Baltimore, Maryland

Professor Lawrence Fredrick
SSE Secretary Emeritus
University of Virginia, Charlottesville, Virginia

Professor Henry Bauer, *JSE* Editor Emeritus
Dean Emeritus, Virginia Tech
Blacksburg, Virginia

Index of Previous Articles in the *Journal of Scientific Exploration*

JOURNAL OF SCIENTIFIC EXPLORATION

ORDER FORM FOR 2008 – 2011 PRINT ISSUES

JSE Volumes 1–21 are available FREE at scientificexploration.org (JSE Vol. 22–current are free to SSE members at http://journalofscientificexploration.org/index.php/jse/login). JSE Print Volumes 22–current (2008–current) are available for $20/issue; use form below or go to www.scientificexploration.org and use the Donation button to pay and order a Journal issue.

Year	Volume	Issue	Price	×	Quantity	= Amount
			$20			

Postage is included in the fee.

ENCLOSED

Name

Address

Phone/Fax

Send to: SSE
151 Petaluma Blvd. S. #227
Petaluma CA 94952

Fax (1) (707) 559-5030
EricksonEditorial@gmail.com
Phone (1) (415) 435-1604

☐ I am enclosing a check or money order

☐ Please charge my credit card as follows: ☐ VISA ☐ MASTERCARD

Card Number

Expiration

Signature

☐ Send me information on the Society for Scientific Exploration

GIFT JSE ISSUES AND GIFT SSE MEMBERSHIPS

Single Issue: A single copy of this issue can be purchased.
For back issues, see order form on previous page and Index.

Price: $20.00

Subscription: To send a gift subscription, fill out the form below.

Price: $85 (online)/$145(print & online) for an individual
$165/$225 (online only/print) for library/institution/business

Gift Recipient Name ____

Gift Recipient Address _______________________________________

____

Email Address of Recipient for publications_____________________________

☐ I wish to purchase a single copy of the current JSE issue.

☐ I wish to remain anonymous to the gift recipient.

☐ I wish to give a gift subscription to a library chosen by SSE.

☐ I wish to give myself a subscription.

☐ I wish to join SSE as an Associate Member for $85/yr or $145/yr (print), and receive this
quarterly journal (plus the EdgeScience magazine and WISE newsletter).

Your Name ____

Your Address ____

____

Send this form to: *Journal of Scientific Exploration*
Society for Scientific Exploration
151 Petaluma Blvd. S., #227
Petaluma CA 94952 USA

Fax (1) (707) 559-5030
EricksonEditorial@gmail.com
Phone (1) (415) 435-1604

For more information about the Journal *and the Society, go to*
http://www.scientificexploration.org

JOIN THE SOCIETY AS A MEMBER

The Society for Scientific Exploration has four member types:

Associate Member ($85/year with online Journal; $145 includes print Journal): Anyone who supports the goals of the Society is welcome. No application material is required.

Student Member ($40/year includes online Journal; $100/year includes print Journal): Send proof of enrollment in an accredited educational institution.

Full Member ($125/year for online Journal; $185/year includes Print Journal): Full Members may vote in SSE elections, hold office in SSE, and present papers at annual conferences. Full Members are scientists or other scholars who have an established reputation in their field of study. Most Full Members have: a) doctoral degree or equivalent; b) appointment at a university, college, or other research institution; and c) a record of publication in the traditional scholarly literature. Application material required: 1) Your curriculum vitae; 2) Bibliography of your publications; 2) Supporting statement by a Full SSE member; or the name of Full SSE member we may contact for support of your application. Send application materials to SSE Secretary Mark Urban-Lurain, urban@msu.edu

Emeritus Member ($85/year with online Journal; $145/year includes print Journal): Full Members who are now retired persons may apply for Emeritus Status. Please send birth year, retirement year, and institution/company retired from.

All SSE members receive: online quarterly *Journal of Scientific Exploration* (*JSE*), *EdgeScience* online magazine, WISE online newsletter, notices of conferences, access to SSE online services, searchable recent *Journal* articles from 2008–current on the JSE site by member password, and for all 25 years of previous articles on the SSE site. For new benefits, see WISE, the World Institute for Scientific Exploration, instituteforscientificexploration.org

Your Name ____

____

Email____

Phone________________________ Fax________________________

Payment of ___________________ enclosed, *or*

Charge My   VISA ☐   Mastercard ☐

Card Number ______________________ Expiration________

Send this form to:
Society for Scientific Exploration
151 Petaluma Blvd. S., #227
Petaluma CA 94952 USA

Fax (1) (707) 559-5030
EricksonEditorial@gmail.com
Phone (1) (415) 435-1604

For more information about the Journal *and the Society, go to*
http://www.scientificexploration.org

JOURNAL OF SCIENTIFIC EXPLORATION
A Publication of the Society for Scientific Exploration

Instructions to Authors
(Revised March 2010)

All correspondence and submissions should be directed to:

JSE Managing Editor, EricksonEditorial@gmail.com, 151 Petaluma Blvd. So., #227, Petaluma CA 94952 USA, (1) 415/435-1604, *fax* (1) 707/559-5030

Please submit all manuscripts at http://journalofscientificexploration.org/index.php/jse/login (please note that "www" is NOT used in this address). This website provides directions for author registration and online submission of manuscripts. Full author's instructions are posted on the Society for Scientific Exploration's website, at http://www.scientificexploration.org/documents/instructions_for_authors.pdf for submission of items for publication in the *Journal of Scientific Exploration*. Before you submit a paper, please familiarize yourself with the *Journal* by reading articles from a recent issue. (Back issues can be browsed in electronic form with SSE membership login at http://journalofscientificexploration.org, click on Archive link; issues before 2008 are freely accessible at http://www.scientificexploration.org/journal/articles.html) Electronic files of text, tables, and figures at resolution of a minimum of 300 dpi (TIF or PDF preferred) will be required for online submission. You will also need to attest to a statement online that the article has not been previously published and that it is not currently submitted elsewhere for publication.

AIMS AND SCOPE: The *Journal of Scientific Exploration* publishes material consistent with the Society's mission: to provide a professional forum for critical discussion of topics that are for various reasons ignored or studied inadequately within mainstream science, and to promote improved understanding of social and intellectual factors that limit the scope of scientific inquiry. Topics of interest cover a wide spectrum, ranging from apparent anomalies in well-established disciplines to paradoxical phenomena that seem to belong to no established discipline, as well as philosophical issues about the connections among disciplines. The *Journal* publishes research articles, review articles, essays, book reviews, and letters or commentaries pertaining to previously published material.

REFEREEING: Manuscripts will be sent to one or more referees at the discretion of the Editor-in-Chief. Reviewers are given the option of providing an anonymous report or a signed report.

In established disciplines, concordance with accepted disciplinary paradigms is the chief guide in evaluating material for scholarly publication. On many of the matters of interest to the Society for Scientific Exploration, however, consensus does not prevail. Therefore the *Journal of Scientific Exploration* necessarily publishes claimed observations and proffered explanations that will seem more speculative or less plausible than those appearing in some mainstream disciplinary journals. Nevertheless, those observations and explanations must conform to rigorous standards of observational techniques and logical argument.

If publication is deemed warranted but there remain points of disagreement between authors and referee(s), the reviewer(s) may be given the option of having their opinion(s) published along with the article, subject to the Editor-in-Chief's judgment as to length, wording, and the like. The publication of such critical reviews is intended to encourage debate and discussion of controversial issues, since such debate and discussion offer the only path toward eventual resolution and consensus.

LETTERS TO THE EDITOR intended for publication should be clearly identified as such. They should be directed strictly to the point at issue, as concisely as possible, and will be published, possibly in edited form, at the discretion of the Editor-in-Chief.

PROOFS AND AUTHOR COPIES: Authors will receipt copyedited, typeset page proofs for review. Print copies of the published Journal will be sent to all named authors.

COPYRIGHT: Authors retain copyright to their writings. However, when an article has been submitted to the *Journal of Scientific Exploration* for consideration, the *Journal* holds first serial (periodical) publication rights. Additionally, after acceptance and publication, the Society has the right to post the article on the Internet and to make it available via electronic as well as print subscription. The material must not appear anywhere else (including on an Internet website) until it has been published by the *Journal* (or rejected for publication). After publication in the *Journal*, authors may use the material as they wish but should make appropriate reference to the prior publication in the *Journal*. For example: "Reprinted from [or From] "[title of article]", *Journal of Scientific Exploration*, vol. [xx], no. [xx], pp. [xx], published by the Society for Scientific Exploration, http://www.scientificexploration.org."

DISCLAIMER: While every effort is made by the Publisher, Editors, and Editorial Board to see that no inaccurate or misleading data, opinion, or statement appears in this *Journal*, they wish to point out that the data and opinions appearing in the articles and announcements herein are the sole responsibility of the contributor concerned. The Publisher, Editors, Editorial Board, and their respective employees, officers, and agents accept no responsibility or liability for the consequences of any such inaccurate or misleading data, opinion, or statement.

CPSIA information can be obtained at www.ICGtesting.com
Printed in the USA
BVOW011503151212

308271BV00006B/162/P